SpringerBriefs in Geography

SpringerBriefs in Geography presents concise summaries of cutting-edge research and practical applications across the fields of physical, environmental and human geography. It publishes compact refereed monographs under the editorial supervision of an international advisory board with the aim to publish 8 to 12 weeks after acceptance. Volumes are compact, 50 to 125 pages, with a clear focus. The series covers a range of content from professional to academic such as: timely reports of state-of-the art analytical techniques, bridges between new research results, snapshots of hot and/or emerging topics, elaborated thesis, literature reviews, and in-depth case studies.

The scope of the series spans the entire field of geography, with a view to significantly advance research. The character of the series is international and multidisciplinary and will include research areas such as: GIS/cartography, remote sensing, geographical education, geospatial analysis, techniques and modeling, landscape/regional and urban planning, economic geography, housing and the built environment, and quantitative geography. Volumes in this series may analyze past, present and/or future trends, as well as their determinants and consequences. Both solicited and unsolicited manuscripts are considered for publication in this series.

SpringerBriefs in Geography will be of interest to a wide range of individuals with interests in physical, environmental and human geography as well as for researchers from allied disciplines.

More information about this series at http://www.springer.com/series/10050

Frederic R. Siegel

Cities and Mega-Cities

Problems and Solution Strategies

Springer

Frederic R. Siegel
George Washington University
Washington, DC, USA

ISSN 2211-4165 ISSN 2211-4173 (electronic)
SpringerBriefs in Geography
ISBN 978-3-319-93165-4 ISBN 978-3-319-93166-1 (eBook)
https://doi.org/10.1007/978-3-319-93166-1

Library of Congress Control Number: 2018944371

This Springer imprint is published by the registered company Springer International Publishing AG part of Springer Nature.
The registered company address is: Gewerbestrasse 11, 6330 Cham, Switzerland

I dedicate this book to my daughters Gabriela and Galia Siegel and my grandchildren Naomi, Coby, and Noa Benveniste and Solomon and Beatrice Gold. I hope for them that solutions to global socio-environmental problems are supported economically by institutions and countries with strong monetary reserves and technological expertise to better the quality of life for all Earth's inhabitants in the 2035 and 2050 worlds and beyond.

Prologue

Creating a new purpose-built city from scratch that will be sustainable and provide a good quality of life to all its future inhabitants can be done with thorough land-use and city planning by a team of experienced experts who interact openly with each other. This has been the case in the latter half of the twentieth century with the construction and functioning of the "new" capital cities of Brasilia, Brazil; Abuja, Nigeria; and Astana, Kazakhstan. The challenge is far greater when planning the redevelopment/rehabilitation of existing, densely populated cities that have allowed shantytowns/slums to spring up and grow sometimes within a city but more often at the perimeter skirting a city. This is the case mainly in less developed and developing countries in Asia and Africa with some in South America. A shantytown or slum is defined in this text as a densely populated (crowded), dirty, run-down, squalid part of a city consisting of a large number of crudely built dwellings inhabited by poor people often underserved with the basic necessities of life. As I write at the end of 2017 and beginning of 2018, ~4 billion (~54%) of the 7.5+ billion global population live in cities, but in 2050 this urban population number will reach 6.9 billion (~70%) of the projected 9.8 billion global population. In 2014, 881 million of the 7.2 billion people on Earth (~12%) lived in these downtrodden locations, most often underserved, for example, by not having access to safe water, adequate sanitation, well-built shelter, and health clinics, among others, of the sustainable quality of life needs other city inhabitants received. The number of shantytown/slum inhabitants in 2017 is likely over 1 billion. Residents of the shantytowns/slums are neither protected from floods nor from other natural or anthropogenic hazards that might impact their locations. The question then is how can a national government and a municipality working together with planning teams use available funds to best redevelop/rehabilitate established cities that have such problems and thus improve the quality of life for all city dwellers. What should be done will be obvious to planning teams. What could be done with available funds and funds that would be granted in the future if initial projects show success means prioritization by importance. This begins with clean water, nutritious food, and access to sanitation, the essences of a healthy life. From here we move to redevelop/rehabilitate down a prioritization list. This environmental brief discusses the problems faced by populations in urban environments and their possible solutions whether in mega-cities (37 in 2017 with >10 million people) or other high-population urban centers (482 with >1 million people and another 515 with >500,000 inhabitants) [1]. Subjects may be considered in different sections in different contexts that discuss new city planning and rehabilitation/redevelopment of existing highly populated cities. An example is water for domestic use, for hygiene, for sanitation, its role in subsidence, flooding, as carrier of pollution, and in global warming and extreme weather. It is clear that actions must be taken globally in the immediate future to solve the problems of residents of shantytowns/slums within or proximate to major urban centers lest social disorder affect all a city/country populace.

Reference

1. Demographia World Urban Areas. (2017). 13th Annual Edition. demografia.com/db-worldua-index.htm

Contents

Chapter 1
Introduction

1.1 Demographic Realities

A great demographic shift is taking place as rural populations are moving to urban centers (in-migration) with the hopes of gaining employment, having access to healthcare for their families and education for their children. In mid-2017, of the 7.5+ billion global population, 4.0 billion people lived in cities and 3.5 billion lived in rural settings. By 2050, the projected global population is projected to be 9.8 billion earthlings, with a continuing demographic shift that is estimated to increase the urban population to 6.9 billion leaving a rural population of 2.9 billion people [1]. Some of the urban growth will be natural, much is expected to come from in-migration as noted above, and part will be from immigration as allowed by national laws.

By 2017, there were 519 cities with more than one million inhabitants. Of these, 37 were mega-cities with populations of more than 10 million people, 84 had more than 5 million people [2]. Both urban categories and others will increase during the timespan to 2050. In many of these urban centers and those with smaller populations (e.g., 100,000 to 500,00), delivery of "essential" services such as access to safe drinking water and to adequate sanitation, as well as to healthcare and utilities (e.g., gas, electricity [in secure housing]), has not extended to areas of rapid, ofttimes uncontrolled and/or unregulated peripheral growth. The result has been population vulnerability to water-borne diseases (e.g., diarrhea) and vector-borne diseases (e.g., malaria, dengue fever) where protective measures have not been put in place or for which there are no vaccines, vaccines being reevaluated (e.g., for dengue fever), or where available vaccines have not been applied because of misinformation, religious belief, or terrorizing actions by insurgent groups. Populations are vulnerable as well to natural hazards that have historically impacted urban areas: earthquakes, flooding, landslides, extreme weather (e.g., hurricanes and storm surges, drought, heat waves). Pollution can be a threat to human health from

F. R. Siegel, *Cities and Mega-Cities*, SpringerBriefs in Geography,
https://doi.org/10.1007/978-3-319-93166-1_1

industrial development proximate to urban populations and weather conditions. An example is long lasting smog and air polluted with <2.5 μ particulates that access deep into the lungs and that is estimated to contribute to 3.3 million premature deaths annually [3]. This figure could double by 2050 if sources of this outdoor air pollution are not required by law with enforcement teeth to use the best available technology to capture or otherwise significantly reduce their emissions. In a 2016 report, the United Nations presented the status of sustainability for global populations, especially those in underserved urban and rural locations [4].

1.2 Urban Planning Complexities

Land-use planning is a complex, long term, ongoing process that is based on truly multi-discipline sets of principles that must complement each other at all levels of decision making. It is complex when planning a new city that is purpose built. The building of a new city may be to attract population away from densely populated cities/mega-cities and relieve that population pressure there (e.g., from Beijing, active, from Cairo, planned, from Mexico City, being contemplated), and perhaps from multiple mega-cities in India (e.g., Delhi, Mumbai, in consideration). Building of a new city has also been purposed to spread out development possibilities to lightly populated regions (e.g., Brasilia, functioning). It may be to build a new capital city, Abuja (functioning), in an interior location to escape a coastal capital city (Lagos) with grave environmental problems and with much squalor and with large underserved populations living in shantytowns (~8 million or more of the more than 20 million population). A journal article on future cities and environmental sustainability lists 10 factors that global urbanization has to consider and their mitigation possibilities based in good part on technological innovations, good management, economic stability, and inclusive urban development team planning. These include coping with declining per capita water resources by water purification and rainwater harvesting, recycling of much of the 70% of trash that is not recycled, dealing with increased energy needs by use of renewable sources, countering urban (heat island) warming by using reflective building materials, preserving or opening green spaces, and increasing biodiversity by creating protected ecosystems [5]. However, this article lacks a 'down to earth' consideration of the geological, topographical, pedological features that are basic to planning future purpose built cities or rehabilitation/redevelopment of cities with greatly underserved populations. Rehabilitation/redevelopment is certainly more complex when planning for existing cities especially those with high populations, mainly in Asia and Africa, that include large shantytown/slum populations (e.g., with lack of safe water, proper sanitation, reasonable shelter, adequate health facilities, electricity, good transportation infrastructure, among other problems). This has to be accomplished in well planned, prioritized stages as is underway in India's smart cities project described later in this chapter.

I am a geologist and assert that land-use planning must start with a knowledge of the soils and their characteristics and of surface and subsurface rock types in an area, the rock structure (e.g., if sedimentary, their attitude such as horizontal, dipping or angled) and other characteristics (e.g., mineral composition, degree of lithification [solidification]). In addition, the topography (e.g., steep relief, gently hilly or not), history of or observable and/or measurable signs of past hazard events (e.g., of earthquakes, floods, landslides/mudflows) fall in the realm of a geologic evaluation when laying the foundations for a land-use planning team. Finally, a geologist has to assess the susceptibility of an area location to extreme weather events such as high energy tropical storms, extended droughts, and heat waves that are likely to increase in the future as a result of global warming. This and many other interrelated disciplines will be discussed in the text as multidisciplinary teams produce plans for city development/redevelopment from this geological, topographical, and pedological foundation.

Municipalities design master plans to guide land-use planners in their assessments of parcels that are in line for development. Done correctly, this effectively dictates what development can take place and assures that each land use phase melds seamlessly with planned use of juxtaposed and nearby parcels. Nonetheless, planners have to be able to adapt a preferred use of terrain to reflect changes, especially those driven by population growth and government decisions based on perceived need and competition for development.

1.3 Planning Guides

There are several advisories that should be followed to assure maximum benefits from land-use planning. These include:

1. Manage development of terrain so as to meet societal needs and protect environments and their contained ecosystems in an effective, efficient, and ethical manner. This means making systematic analyses in all facets during the planning process, and assessing viable alternatives for land-use options that serves the best interests of society within economic and political restraints.
2. Meet the needs of populations by protecting the sustainability of natural resources for the foreseeable future by minimizing environmental intrusions on support ecosystems and allowing only development that is sustainable.
3. Regulate land uses, (residential, commercial, light industry, heavy industry, waste disposal, and agricultural) so that they interrelate, not one to another but all to all others. Regulations will determine disposition, spacing, and utilization of terrain for development that takes into account the best interests of the community including inhabitants in peripheral shantytown/slum environments.
4. Plan for future population growth and development that are sustainable and meet societal needs. This means being able to adapt to changes that may alter land-use plans because of changes issuing from existing rural and urban physical, social,

economic, and political conditions. There may be no mandate to do so but decisions should be made with citizen expert input to planning team evaluations. Ultimately, government decision makers have to show political will given socioeconomic realities based on a land-use plan that has been prepared according to the above propositions. These and other criteria for optimum benefits to populations and earth systems from well thought out and implemented land-use planning are discussed in this book.

1.4 Brasilia, a Purpose Planned City Adapted to a Population Overrun

Lastly, there are two scenarios (one partitioned) that land-use planning teams have as challenges. First, is where land-use planning was successfully applied to build a modern city from the start, Brasilia City. Brasilia City, in the interior of Brazil, was purpose built to be the center of government and with the aim of spreading development from the country's coast to the interior as measured by the per capita Gross Domestic Product (GDP). It spread the population the but not the per capita GDP except locally for Brasilia [6]. City construction began in 1957 and was completed in 1960. The existing population in the region selected for Brasilia was estimated to be 64,000 people with the original plan calling for a permanent population of 500,000. However, by 2016, the population in the center city and adjoining neighborhoods was greater than 2.5 million people but with a low population density [7]. Agricultural development in the surrounding area contributes to the metropolitan area food supply. Adaptation of the land-use planning expansion norms has been able to accommodate the increased population with clean water, sanitation systems, and other needs to sustain the Brasilia region. Water comes from a reservoirs created by the damming of two rivers that flow to treatment plants before its distribution. As water demand increases, the water supply will be augmented by water from manmade Lake Paranoá [8].

The second scenario that is more challenging to overcome is applying land-use planning experience in the rehabilitation of existing cities. This has two approaches defined largely by the economic investment a country is able to put into the rehabilitation processes via self financing, foreign investment, loans or grants and by the results one hopes to achieve when finances are committed. First is the identification of problems that require rehabilitation/redevelopment and will help people that need it most. Second is the prioritization of where best to invest as a start and subsequently followup. In many highly populated cities, the main problem is one that improves public health and involves water: the need for a clean water supply and access to adequate sanitation.

Investment to attack a public health problem can be rewarding socially and economically. For example, the Bangladesh government made an investment in public health that gave stupendous results. In 2003, 42% of the Bangladeshi population practiced open defecation. Over the next 12 year, the Bangladeshi government invested 25% of national development funds in education and engineering with the result that by 2016 only 1% of the population followed the unsanitary practice. In addition, improved sanitation coverage was at 61% versus 28% in 2003 [9]. This greatly mitigated the incidence of diarrheal diseases from use of polluted water and diminished the conditions that could support the start of an infectious disease epidemic (e.g., cholera, vis a vis Haiti, Yemen). It also improved the per capita income from $1154 in 2012–2013 to $1514 in 2014–2015, a 31.2% increase.

1.5 India's Smart City Project: Rehabilitation That Serves All Citizens

India has initiated a smart cities project that is not a complete land-use planning makeover but rather focuses on improving public health condition for citizens and on improving populations' living conditions [10]. The focus is on five major areas for improvement. These include two, water supply and sanitation/waste management that, in theory, and in time will improve public health conditions for 100 s of millions of Indian citizens. In 2015, 48% of the Indian population (~600 million people) practiced open defecation. The smart city project will be a start in bringing this number down. It promises 24/7 direct (safe) water supply connections and calls for 100% collection of all water charges (to support the water supply network). The sanitation and waste disposal phase of the project proposes to provide all households with access to toilets connected to a waste water network or a daily doorstep collection system (where there is no sewer transfer to a collection and treatment facility…but no backyard or street disposal). Schools will have separate toilets for girls and all toilets will have hand washing stations. In addition the smart cities project proposes to provide electricity 24/7 for all households together with telephone and mobile connections with 100% Wi-Fi coverage throughout a city. For spatial planning, a city would have 95% of its residents within 400 m of primary schools, retail sites, parks, and other recreational areas. For the transport part of the smart cities plan, there would be a 30 min maximum travel time to work for small and medium cities and 45 min for metropolitan areas. There would be safe footpaths and bicycle lanes on all streets in both directions. Importantly, there would be high quality and high frequency public transportation within 800 m of all residents where there is a population density of 175 persons per hectare. The Modi smart cities project is ambitious with proposals from 100 cities evaluated for a pilot program start. The project will be costly, but if completed would be a worthwhile effort, not withstanding other needed rehabilitation steps such as reducing air pollution, protection against flooding and landslides, and reasonably accessible staffed and supplied healthcare facilities.

References

1. Population Reference Bureau. (2017). World population data sheet 2017, Washington, DC.
2. Demographia world urban areas. (2017). 13th annual edition. demografia.com/db-worldua-index.htm
3. Lelieveld, J., Evans, J. S., Fnais, M., Gianniadaki, D., & Pozzer, A. (2015). The contribution of outdoor air pollution sources to premature mortality on a global scale. *Nature, 525*, 367–371. https://doi.org/10.1038/nature15371.
4. United Nations. (2016). The sustainable development goals report 2016. New York, 52p. Online. www.un.org/lb/Library/Assets/The-Sustainable-Development-Goals-Report-2016-Global.pdf
5. Pifat, S., Powell, R., & Aydin, D. (2016). Future cities and environmental sustainability. Online. https://doi.org/10.1186/s40984-016-0014-2.
6. Grimes, A., Matlaba, V. J., & Poot, J. (2017). Spatial impacts of the creation of Brasilia: A natural experiment. *Environment and Planning A, 49*, 784–800.
7. Brasília. http://www.en.wikipedia.org/wiki/Bras%C3%ADlia
8. IWRM. (2010). Water for the future of Brasilia. Online. www.bmbf.wasserresourcen-management.de/en/554.php
9. Bhattacharjee, R. (2016). Bangladesh: First South Asian nation to become open defecation free. May 4, South Asian Monitor. Online. southasianmonitor.org/detail.php/type=sl$nid=16770
10. Ministry of Urban Development, India. (2015). Smart City – Mission statement and guidelines. 43p, (18 text, 25 annexures).

Chapter 2
Foundations for a Land-Use Planning Project Determine Sustainability

2.1 Foundations for Land-Use Planning

Land-use planning is multi-faceted and complex whether designing a new city from its start or adapting a municipality for expansion to accommodate an increasing population. It becomes more complex to plan the rehabilitation/redevelopment for a high population city to redress the increasing population problem when a government is not providing the basic needs (e.g., clean water, sanitation, proper shelter) for inhabitants of unplanned shantytowns or slums bordering a city that a municipality allowed to grow to densely populated areas. The likely happened because the shantytown/slum population provided cheap labor for development projects. Land-use planning for the new capitol city Brasilia, as described in Chap. 1, is an example of thorough, successful venture that planned for accommodating natural population growth and in-migration from other states in the country [1]. The starting point in any project of land-use and city planning has to begin with siting in the area being assessed for development.

As emphasized in Sect. 1.2, there are three foundations for site evaluation for a new city or when gauging rehabilitation/redevelopment of an existing city. These are the geology, topography, and soils of the location(s) under study. Geologically, it is necessary to know the rock types, their structure, and unique characteristics whether exposed at the surface or underlying soils or in the deeper subsurface and the geological hazard history. Topographically, the geologist evaluates valleys and lowlands, and maps their floodplains as well as categorizing hilly and steep terrain as stable or not with respect to landslides. Soil characteristics define how they will respond to construction on or in them, or their response to events such as rainfall or tremors, or if they should be reserved for parks, or for near city agriculture as truck farms.

F. R. Siegel, *Cities and Mega-Cities*, SpringerBriefs in Geography,
https://doi.org/10.1007/978-3-319-93166-1_2

2.2 Geology

We will first discuss geology and follow this with how geology ties in with topography and soils in land-use planning. A geology team prepares maps that show the disposition of rock types in the study area. As many of us learned in middle school or in an introductory geology course in college, there are three major classes of rocks: igneous, sedimentary, and metamorphic. Each of these has many subclasses, all of which have specific properties such as load bearing strength, response to earthquake motion, susceptibility to landslides, and a possibility of being host to an aquifer or to other natural resources. This helps define how geology affects a decision on best use of a terrain. In addition, a geology team describes the structures the rocks have such as angles they present with respect to the horizontal plane, folds in the rock, and fractures and fault ruptures that may indicate past earth movements. The structures are indicated by symbols on a geologic map or may be shown on cross sections (profiles of rocks underlying the study area – a slice into the earth) [2]. Analyses of the rocks at the surface and those in the subsurface (revealed by cores) reviewed by a geology team define the properties that can affect site areas for land-use planning such as noted earlier. In addition, rock chemistry and possible gas emissions may put limitations how land may be used.

2.3 Topography

Topographic analysis by geologists and geographers complements geology by determining best use advice for parcels of land. It is presented to city/land-use planning teams as a map. Topographic maps, often light green in color in vegetated areas, show cultural features (e.g., structures, roads, railroads, power lines). They show streams, river, lakes, ponds, and seas/oceans in blue. Most important are the lines on them called contours. Contour lines mark contiguous points at the earth's surface with the same elevations above sea level. Contour spacing may represent detailed elevations (e.g., 10 ft intervals for flatter areas or 100 ft intervals for high relief elevation spacing). A study of the contours on a topographic map delineates flat areas (e.g., plains, mesas), low hilly areas, areas with steep hills and cliffs, river valleys and flood plains, areas with depressions or sinkholes and perhaps disappearing streams or rivers [3]. When coupled with a geologic map, a planning team can put limitations of how parcels of land can best be used. For example, a hilly area underlain by loosely consolidated sedimentary rocks that slope towards a low area is at risk of landslides. Indeed, a geological field survey can often delineate areas that have had landslides by finding scars on the terrain called scarps and other revealing features. Whether or not a hillside is susceptible to landslides is a function partly of the strength and other physical/chemical properties of rocks as well as climate or the likelihood of an earthquake event. An earthquake's shaking, jarring, rolling motions can trigger a landslide. In an area that receives sustained rainfall or torrential rainfall, the rain water can seep into the sedimentary rocks because of their properties of porosity and permeability, add weight to a slope, and lubricate the rocks so as to decrease friction between layers of rock causing them to yield to the pull of gravity and result in a landslide or a mudflow. Depending on a natural hazard event/climate history, use of such land could be off limits for habitation.

2.4 Soils

Soils are the third surface/subsurface feature to be studied in parcels being evaluated by land-use planning teams for new development or redevelopment. There are several classes of soils that form from the disintegration and decomposition of underlying rocks as a function of geography/climate (temperature, precipitation). Soils often carry an imprint of the chemistry of the rocks from which they originated. Soil analyses determine their bearing strength and compressibility, their chemical composition, their suitability for sustainable agriculture, and their response to earthquake motion (thixotropy) and reaction to extreme weather such as by wetting and expansion from precipitation or shrinkage from droughts [4].

The information garnered by geology, topography, and soils is the starting point in the evaluation of land being studied for creation of a city or a population center and its neighborhoods being assessed for rehabilitation/redevelopment. This could be for residential and commercial use, manufacturing, light industry, heavy industry, and waste disposal, and the infrastructure needed for all sectors to function effectively and efficiently.

2.5 Water Resources, Keys to Development/Redevelopment

Key assessments for sustainable land-use options have to include water availability for drinking, cooking, personal hygiene, and sanitation (sewer systems) in residential and mixed use areas for the maximum projected population that could occupy the land. Water must be available to service commercial interests, manufacturing, and light and/or heavy industry that may be in development plans for a municipality. To this must be added water to irrigate food crops and serve animal husbandry in order to insure food security if there are agricultural ventures that draw water from the same source(s) as urban ventures. From an economic perspective, the World Bank estimates that water scarcity exacerbated by climate change could cost some regions such as Central Africa and East Asia as well as the Middle East and the Sahel in Northern Africa as much as 6% of their GDP growth rates by 2050 because of its impacts on agriculture, health and incomes [5]. The World Bank suggests that better water resource management practices could reverse the possible GDP decline. Without sufficient water neither economic-social development nor redevelopment can succeed. Hydrogeologists, water engineers, sanitary engineers, and water chemists are city/land-use planning team members that assess existing and future water availability for populations and development plus water quality for the purposes cited above.

Water resources **to serve the long term needs** of communities can come from aquifers in the shallow or deep sub-surface if they are recharged at the same rate they are being discharged. If this is not the case, an aquifer will eventually be depleted and possibly cause subsidence of the overlying surface with the problems subsidence can cause (described in Chap. 3). Water from rivers, sometimes dammed to provide reservoirs and sometimes rerouted, and from lakes are major sources of water for communities. A latter sourcing, such as rerouting rivers to serve agricultural projects has caused severe problems for people and sensitive ecosystems in the former Soviet Union and China. The use of water from the Logone river/Lake Chad system bordering Chad, Niger, Nigeria, and Cameroon, for irrigation to sustain agriculture has caused the lake to shrink from being the largest fresh water lake in Africa to ~5% of its original size (from ~the size of the state of Vermont in the United States to the ~size of the state of Rhode Island).

2.5.1 Water Sources That Sustain Citizenry

There are two types of aquifers designated as unconfined and confined. They may be relatively close to the surface or very deep in the sub-surface. A confined aquifer may contain huge amounts of water that may or may not be potable. Because it is sealed off from recharge, its use as an aquifer lasts only until the water in it is depleted. Conversely, an unconfined aquifer can recharge with rain water, snowmelt, or inflow from rivers it intersects. If the discharge of aquifer water is continually greater than recharge, the aquifer water table (level of water in the aquifer) will drop, a condition that not only depletes available water and can ultimately lose this source but can also cause surface subsidence with problems that will be considered later in this section. Where recharge equals discharge, aquifer utility can be long lasting while there is no marked change in precipitation in recharge zones from global warming/climate change.

Rivers and to some degree streams provide water supplies for urban and rural communities. The quality of the water as determined by chemical and biological analyses dictates the degree of treatment water may need before being acceptable for drinking or other uses in a municipality's planning scheme. The volume of water a river moves varies from season to season depending on rainfall, snow/ice melt, or drought but flows are generally close to average although there can be periods of greater than average precipitation that can cause minor and major flooding. Cape Town, South Africa, with a population of 3.7 million people, is suffering from severe drought as I write in March, 2018. The main water supply is river water held in reservoirs that are slightly above 1/4 full with no meaningful rain in the forecast. Residents are rationed 50 L (0.05 m^3) a day for all their needs with a projected zero day when no water will flow totals as August 27, 2018. To prepare for future such extreme weather, the municipality is considering installation of a desalination plant, exploitation of aquifers, and recycling of waste water. There is concern about the effect of climate change on surface and aquifer water supplies such as shifts in precipitation volumes and/or patterns. This will be discussed later in this section.

Natural lakes and lakes formed by damming rivers provide water supplies to many municipalities world wide. The number of people that can be serviced by water from lakes depends on the water body volume and discharge/recharge relation. Before use, chemical and biological analyses establish the water quality and need for treatment to remove toxins before releasing water for human or other uses. As with aquifers and streams/rivers, when recharge equals discharge, lakes can provide a steady water supply. Conversely, a lake can shrink as happened in Africa's Lake Chad described in a previous paragraph.

Desalination plants sustain large populations in water-poor global sites such as in the Arabian Peninsula and where aquifers, river waters, or lakes as may exist in coastal regions can not provide enough water as populations grow through natural birth rate and presence of guest worker populations. Desalination plants are expensive to operate because of the energy requirements that drive the distillation processes. Hence, use in economically disadvantaged states or nations that have water needs and access to the seawater or briny groundwater feedstock is generally not feasible. Table 2.1 shows how per capita annual water availability will severely change to lesser volumes in many countries as a result of growing national populations, thus threatening their long term sustainability [6–8]. In addition, recycling of tainted sewer water through treatment facilities is a method being considered for use by a growing number cities today to extend their water supplies.

Although not applicable for supplying potable water to large populations, it should be noted that recently (2016), researchers at Nanjing University have described a system based on a solar still design used since ancient times. That system was inefficient in using solar heat to evaporate seawater from a central pool that condensed on slanted glass roofs and rolled down into collector channels that delivered the water to storage receptacles. The researchers modified the design and used 1 m^2 aluminum foil sections with an array of 300 nanometer wide perforations (0.0003 mm or 0.3 μ) that they misted with aluminum oxide and floated on a sea water pool covered by a slanted roof of glass or plastic sheeting. The system was 4× more efficient in focusing the a broad spectrum of absorbed solar (heat) energy than the sun alone. Each square meter of floating aluminum foil produced **2–8 L of water an hour**, depending on the sun exposure, with salt level below WHO standards for drinking water. Each floating island of foil could be used for 25 cycles [9]. This method of desalination could supplement water supplies for water poor areas near the oceans or with access to briny aquifer waters. The extracted salt can be sold to refining or chemical companies.

Table 2.1 Examples of countries where growing populations reduce per capita renewable water availability

Country	Actual renewable water resources km^3	Population Mid-2014 × 10^6	Per capita water 2014 m^3	Population Mid-2035 × 10^6	Per capita water 2035 m^3
Under grave water stress					
Saudi Arabia	2	30.8	65	37.4	54
Egypt	58	87.9	660	121	479
Morocco	29	33.3	870	38.1	761
Kenya	30	43.2	694	63.7	471
Ethiopia	110	95.9	1147	132.9	828
Somalia	14	10.8	1296	16.5	848
Water stress will shift from marginal to grave					
Nigeria	286	177.3	1613	261.9	1092
Uganda	66	38.8	1701	58.1	1136
Water stress will remain marginal					
India	1897	1296.2	1464	1530	1240
Iran	138	77.4	1783	87.9	1570
Tanzania	91	50.8	1791	81.5	1117
Water condition will shift from sufficient to marginal					
Afghanistan	65	31.3	2077	45.5	1429
Ghana	53	27.0	1963	37.1	1429
Water will remain sufficient					
China	2829	1364.1	2074	1411	2004

Water stress is grave with 1000 m^3 per capita, marginal with 1000–<2000 m^3 per capita, and water sufficient with >2000 m^3 per capita [6–8]

2.5.2 *Effect of Global Warming/Climate Change on Water Resources*

Global warming/climate change will affect water sources that support growing populations, agriculture, and industry in many areas of the world, especially regions in Africa, Asia, South America, and Europe. Some countries in the latter three regions depend to some degree on melt water from alpine glaciers and seasonal snowpack to feed rivers and recharge aquifers. Observations and measurements on alpine glaciers in the Himalayas, Alps, Andes, and Rockies show that mountain glaciers are receding with the exception of the Perito Moreno glacial region of southern Argentina. During glacial ice retreat, the seasonal melt water will continue to flow into rivers and to recharge aquifers. Glacial advance that follows seasonally is basically recharge of a water supply. However, with continued retreat, this source of water to sustain populations will diminish and reduce water supplies thus threatening every day life. Certainly, planners for future new cities or for redevelopment of existing cities have to present options for the future that will maintain water security for the ~2.3 billion more people on earth in 2050 than in 2018, mainly in Africa and

Asia with their burgeoning populations. Planners will have to formulate adaptations to deal with lesser rainfall in the future such as in mid-latitude wet regions if the Intergovernmental Panel on Climate Change are taken into account [10].

In many regions, economically rich and water secure countries will have to contribute financially and technologically to plans that will satisfy water needs for growing population in order to maintain global societal order. This might mean investing in pipelines that can move the water from water rich regions to regions with severe deficits of water. It might mean constructing and maintaining desalination facilities for urbanized coastal regions. It might mean constructing and maintaining recycling facilities that collect and treat dirty water that can be distributed to populations to support their basic needs and the needs that sustain food security (agriculture) as well as manufacturing and industry. The failure to make such investments now would certainly set the table for large scale migration of water refugees and water wars in the not too distant future. This is unthinkable but a real possibility because it is happening in 2018 (e.g., one million mainly South Sudanese in Uganda and 250,000 Somalis in northwest Kenya escaping drought, famine and war), to a seemingly increasing degree. Egypt is in discussions with Ethiopia about Ethiopia's construction of the largest hydroelectric dam in Africa because of the possibility that Egypt's current share of the Nile River waters would be reduced thereby threatening the well-being of the Egyptian population. If this does happen, a conflict between the countries could arise.

In an interesting journal paper, the problem of deficiency of water with global temperature reaching 1.5 °C and 2 °C above the pre-industrial revolution global temperature has been investigated by Chinese scientists and collaborators from the UK, South Korea, Sweden, Switzerland, and the USA. Their results should be of interest to "long-term" city/mega-city planning teams evaluating water supplies for growing urban populations. The scientists studied data from 27 global climate models to identify regions where warming to 1.5 °C and 2 °C above pre-industrial revolution global temperatures could become significantly drier and result in aridification (researchers' term) [11]. This is determined by the ratio of precipitation to potential evapotranspiration. Aridification would negatively impact water supplies, agriculture, and biodiversity, and provide conditions that favor increased occurrences of drought and wildfires. In 2018, the global temperature is 1 °C above the pre-industrial revolution level and about 16% of the earth's land area is desert. These authors project that aridification would affect 20–30% of the earth's land surface if the mean temperature reached 2 °C. However, they estimate that 2/3 of the affected regions would avoid significant aridification if the global warming was kept below 1.5 °C. In the authors' projections, the regions that would benefit most by keeping global warming to 1.5 °C are parts of southeast Asia, and Central America, plus areas where there is increasing drought in the Mediterranean region, southern Africa, and the eastern coast of Australia. Semi-arid areas of Mexico and Brazil that are in the beginning stages of desertification would benefit as well if warming is kept to 1.5 °C [11].

References

1. IWRM. (2010). Water for the future of Brasilia. Online. www.bmbf.wasserresourcen-management.de/en/554.php
2. Graymer, R. W. (1997). How to read a geologic map. Online. www.nature.nps.gov/geology/usgsnps/gmap/gmap1.html
3. Mark. (2013). How to read and interpret topographic maps. Online. soleadventure.com/2013/02/how-to-read-and-intepret-topographic-maps/
4. U.S. Department of Agriculture, Natural Resources Conservation Service. (2006). Soils – Fundamental concepts. Online. http://www.extension.illinois.edu/soil/concepts/concepts.pdf
5. World Bank. (2016). *High and dry: Climate change, water, and the economy*. Washington, DC: World Bank, 55p. Online. https://openknowledge.worldbank.org/handle/10986/23665
6. Population Reference Bureau. (2014). World population data sheet 2014.
7. Food and Agricultural Organization of the United Nations (FAO). (2003). *Review of world water resources by country*. Rome, 97pp.
8. Siegel, F. R. (2015). *Countering 21st century social-environmental threats to growing global populations, Table 2.2 modified*. Springer, 164pp.
9. Zhou, L., Tan, Y., and others, (2016). 3-D self assembly aluminum nano particles for plasmon-enhanced solar desalination. *In Nature Photonics, 10*, 393–398.
10. Intergovernmental Panel on Climate Change. (2014). Climate change 2014. Synthesis report (151pp), Geneva, IPCC.
11. Park, C. E., Jeong, S.-J., & and Others. (2018). Keeping global warming within 1.5°C constrains emergence of aridification. *Nature Climate Change, 48*, 70–74.

Chapter 3
Site Selection with Attention to Susceptibility to Natural and Human Caused Hazards

The site selection for a given use or mixed use urban development project is affected by factors often tied in to an area's geology, topography, and soil conditions. This is a knowledge of the modern and past geologic history of natural hazard events and their triggered hazards that have directly impacted a study area or that occurred at a distance but still caused damage and harmed populations outside the immediate impact area. The frequency, intensity, and duration of a hazard event dictates whether a parcel can be used as is for a given project or projects. If land use is in question or has a must use status, a planning team has to determine if there are mitigation strategies that can be implemented to allow development [1]. These fall in the categories of prediction if possible, prevention if possible, and enhancing the state of preparedness that is definitely possible.

3.1 Natural Hazards

Natural hazards include earthquakes, flooding, landslides, mudflows, extreme weather conditions often intensity enhanced by global warming (drought, heat waves, smog, torrential rains and high velocity winds from hurricanes [typhoons, monsoons]). Add to these a rising sea level in coastal zones, as well as volcanos, wild fires, and occasionally sandstorms. To be complete, the planning team must take into account hazards that can be triggered by a primary event such as tsunami (earthquake), flooding and landslides/mudflows (torrential rains), storm surges (high energy wind-driven storms), and mudflows triggered by melted snow/ice on the slopes of erupting volcanoes. The task of planning against natural hazards seems daunting but doable once the setting is evaluated in itself and in context of the historical and modern hazard events that have been recorded and/or experienced.

The relation between landslides/mudflows, geology, topography, and soil cover is beautifully transparent. Hilly and steep terrain comprised of weakly consolidated

F. R. Siegel, *Cities and Mega-Cities*, SpringerBriefs in Geography,
https://doi.org/10.1007/978-3-319-93166-1_3

sedimentary rocks (comprised of particles not strongly cemented together [lithified]), with or without a thick soil cover, are at risk of landslip especially when they are spatially at an angle that dips to the open face of the terrain. Gravity pulls down on the rocks that are strong enough to resist slides until an event causes movement. The stimulus for a landslip may come from earthquake shaking or jarring motions or from a high rainfall that seeps into the at risk terrain increasing the gravity stress from the weight added by the water and facilitating movement by lubricating contacts between sedimentary rocks or even causing certain sedimentary rocks containing the clay mineral montmorillonite (aka smectite) to swell to further destabilize a hillside. When there is a thick soil cover and torrential rain fall, mudflows and debris slides can result. Planners evaluating use of land proximate to hilly terrain have to assess the risk to projects that may be considered for habitation of what could be vulnerable sites.

Similarly, topography will determine flood vulnerability and potential use of parcels of land in a study area proximate to rivers or streams. Dry areas away from the banks of rivers that fall under water when a river overflows its banks are floodplains. These are categorized according to an estimated interval of recurrence such as a 100 year floodplain (1% chance of recurrence), a 500 year flood plain (0.2% chance of recurrence) and so on. Field geologists make these determinations that provide guidelines for planners surveying land for given projects **but with the caveat that flood recurrence intervals are statistical estimates so that a 100 year flood plain may come under flood waters at a rate greater than a calculated recurrence interval**. This is important for urban planning now but will become increasingly of concern when land-use planning teams assess an area's future potential uses in terrain that is projected to be threatened with more intense, frequent, and longer duration rainfall because of global warming/climate change. Looking to the future, a team can recommend sites for the installation flood control methods (e.g., dams, levees). So that at risk land can be used to serve societal needs.

A knowledge of the geology of an area is essential when evaluating redevelopment in an area that is known to have suffered earthquakes in the geologic past or in modern/recent times. Structures built according to code on strong, solid rocks like granite will have more resistance to earthquake motions than those constructed on less strong sedimentary rocks. Certainly resistance to the shaking, jarring, and/or rolling motions of terrain from an earthquake can be greatly enhanced by stringently following earthquake building codes such as those used in Japan and California (USA) [2, 3]. Geologists can advise land-use planning teams of the rocks present in a study area and characteristics they have that favor earthquake resistance and those that increase risk from earthquake tremors. Recently, California State geologists have found new fault zones in Los Angeles capable of producing magnitude 7 earthquake. These zones have limits on development. A California law mandates that owners of properties in these zones are obligated to hire geologists to ensure that new buildings and major renovations are not located directly on an active fault line. In past California earthquakes, buildings located atop a fault suffered major damage whereas others a few hundred feet away were less affected. The high risk of destruction of existing structures on top of faults has resulted in their demolition such as seven buildings along the San Jacinto fault in the 2000s that were part of San Bernardino Valley College.

As previously noted, earthquakes can trigger landslides by the violently shaking of hilly terrain with a soil cover and underlain by sedimentary rock with a structural angle dipping to a slope face. They can also trigger rock falls in steep topography and otherwise strong rock that may be weakened by fractures or degradation of the rock by weathering processes (e.g., by ice wedging). Earthquakes in the ocean may trigger tsunamis depending on the type of faulting at the earthquake epicenter. As experienced in 2010, a major earthquake in the southwest Pacific Ocean, triggered a tsunami that killed more than 200,000 people mainly in Indonesia but in other countries as well because there was no South Pacific Ocean tsunami warning system that would have allowed an alert so that coastal populations could evacuate inland. A tsunami warning systems is now in place, but more than 200,000 lives too late. Similarly, in 2011, a major earthquake (magnitude 9.0) off of northeast Japan triggered a tsunami that killed at least 19,000 people, ruptured a nuclear power facility (Fukushima Daichi) with the release of radioactivity that caused the evacuation and long-term displacement of hundreds of thousands more from an area of more than a 20 km radius from the facility. Today 7 years after the event, radioactive waters are still leaking into the Pacific Ocean. A thorough geologic study of the coastal area before the building of the nuclear power plant could have revealed that a tsunami that generated waves more than 8 m high impacted the coastal zone in 869 AD and had an estimated recurrence interval of 1000 years [4]. The protective sea wall in the coastal area near the power plant was 5.72 m high. There is little question that if the 5.72 m sea wall had been built up to 8 m or more, the death and damage from the tsunami would likely have been mitigated to a significant degree.

Land-use planning at some distance (within 5–20 mi. [8–32 km]) from active volcanos (erupted during the past 10,000 years) require a geologist's (vulcanologist's) input. Pompeii was 24 km (15 mi.) from Mt. Vesuvius and was wiped out by a glowing ash cloud (nueé ardente) that rolled over the population. Naples with a population of one million people is 21 km (13 mi.) from the volcano. Heat during an eruption of Mt. Rainier in the past melted ice and snow at its summit that mixed with soil and loose rock matter that triggered lahars (hot mud/debris avalanches) that flowed as far as Puget Sound (Oceola, 5000 years ago) and where the town of Ortling is located now (Electron, 500 years ago) [5]. The lahars can travel at more than 20 mph (32 kmph).Today, 10s of thousands of people live in towns built on the deposits of these flows and it is likely that if any land-use planning criteria had been applied to these locations that the towns would have been built elsewhere away from the paths of lahars that reached areas 27 mi. (43 km) away. Warning systems installed and monitored by the U.S. Geological Survey could give people about 1 h to evacuate. Hazard maps for lava flows, tephra and ash falls, as well as mudflows have been prepared for many active volcanoes worldwide. These are essential tools for land-use planning teams to study when they are assessing the usability for population habitation of areas associated within the reach of hazards originating from volcanic activity including warning and alert systems if decisions allow habitation. As I write during January, 2018, the warning and alert system at the erupting Mt. Mayon volcano in the Philippines led to the evacuation of more than 81,000 citizens from at least a 5 mi. (8 km) zone around the volcano because of ash clouds and lava flows. There are no human casualties thus illustrating the success of warning and alert systems. The evacuees are being attended to in prepared safe shelters. There is no sense yet of let up time as the magma mass underlying the volcano continues to swell but authorities are preparing for a 3 month activity period.

A knowledge of soil properties is essential to determine their response to the building new or relocating of residential, commercial and manufacturing/industrial structures and infrastructure. For example, if the load bearing capacity of a soil and underlying rocks is low, it can still be used as long as kinetic equilibrium between a structure and the subsurface is maintained. This can be accomplished by employing a floating foundation strategy that dictates that a weight of earth materials be excavated equal to the weight of the proposed completed structure, workplace accessories, and people that will work there. In assessing sites for expansion development that may be outside initial planned city limits, some areas may have soils best preserved for agriculture that can add to food security.

3.2 Site Planning Against Risk from Anthropogenic and Natural Hazards

Anthropogenic and natural sources of air pollution are considered in two groups: indoor air pollution and outdoor air pollution, both of which contribute to the premature deaths of millions of people annually. In 2015, China estimates that air pollution contributed to the premature deaths of 1.2 million of its citizens. Globally, the World Health Organization gives an estimate of 7 million premature deaths abetted by air pollution during 2012, 3.7 million from indoor air pollution and 3.3 million from outdoor air pollution [6]. Past hazards, their causes, and a community's response to prevent or mitigate the effects of recurrences has to be catalogued. This can alert a land-use evaluation team that may be assessing development/redevelopment possibilities similar to a venue where an anthropogenic or natural hazard has occurred and its effects on people and other life forms in the environment.

Here we can start with pollution…air, water, and soil. A 2017 report by The Lancet Commission on Pollution and Health ascribes 9 million premature deaths annually to pollution (1 in 6 persons die from pollution globally) of which 6.5 million are attributed to air pollution. The economic damage in 2015 was estimated at US$4.6 trillion suffered mainly by less developed and developing countries because of lost productivity that reduced economic output by 1–2% and Gross Domestic Products (GDPs) by up to 2% [7]. Global exposure to air pollution and its disease burden was updated in a 2017 report [8]. Subsequently, we will discuss flooding and dam failures, and the need to thoroughly assess drainage basin changes that may cause flooding or a decrease in water supplies. In addition we will consider the disruption or loss of ecosystems and arable land to human encroachment, failure to install observation and measurement posts and hazard warning and alert systems. Similarly, planning against sabotage by disgruntled employees or by terrorists at industrial facilities that deal with toxic materials including radioactivity sources should include professional staffing and functioning fail safe monitoring systems. Planning at these sites should include double checking to assure that construction strictly adheres to national or local building codes that carry enforcement legislation.

3.2.1 *Air Pollution*

In order to avoid having a useable parcel impacted by air pollution, planning teams must first determine what air pollution problems they may be dealing with (naturally occurring or anthropogenic) such as fine size particulates <2.5 μ, heavy metals, and toxic aerosols and if such contaminants have deposited in the study area in the past (e.g., evidenced by soil analyses). Second, team members must know where and how threatening airborne pollutants originate and the role of a dominant wind direction and other meteorological events in transporting and depositing them.

3.2.1.1 Naturally Occurring

Natural sources of air pollution, especially fine size particulates (<2.5–10 μ) are generally ephemeral and not long lasting. One source would be volcanic ash eruptions. Although the effects of most ash eruptions are short lived, some such as the 1991 eruption of Mt. Pinatubo in the Philippines that cooled the earth ~1 °C (1.8 °F) during the course of a year. Ash eruptions from Eyjafjallajokull volcano in Iceland in 2010 disrupted air traffic in for 8 days in England and a few days or more in other countries in Europe dependent on wind direction and density of the ash being carried in the air flow. Fortunately, such large scale events are not regularly occurring. There are lesser scale ash eruptions in other global locations, such as in Indonesia or the Andes. Planning for any size population center or plans to expand a center to accommodate growing populations has to take into account this source of fine size air pollutants. This means avoiding a new center's location in areas nearby and downwind of an ash eruption that are within its historical reach. Vulnerable existing populations should have warning and alert systems in place to sound a stay indoor advisory, and if necessary, evacuation with routes clearly defined.

Desert sandstorms that carry fine particulates to populated regions are potential hazards if they are unusually long lasting, hurt visibility, and expose people who venture outdoors without masks to fine particle ingestion through inhalation and subsequent bronchial disease. Vehicular urban traffic emissions during a sand storm add to health endangering fine particulate matter. In such cases (e.g., Beijing, 2015), people are warned to stay indoors and traffic advised to pull over before sand gets sucked into a vehicle's engine. Clearly, evaluation of sites for new population centers should take this into account where sandstorms may be a recurring, often seasonal hazard.

Another source of natural air pollution that is non-particulate and not a direct threat to human health is from the flatulent activities, burping, and manure generated by livestock (cattle, pigs, sheep, and goats) that release CH_4 (methane) into the atmosphere. This greenhouse gas adds to others (e.g., CO_2,) generated by human activities that abet the global warming process that favors about climate change. Scientist are working on changes in feed stock that can reduce the volume of methane released [9].

3.2.1.2 Human Activity Caused Indoor Air Pollution: Problem Solutions

Indoor air pollution in existing heavily populated cities originates from the burning of brown (soft) coal, charcoal, and wood in inefficient stoves in poorly ventilated cooking and heating areas of a home. Fine particles and noxious gases released indoors causes sickness and death especially for the very young and the very old. It is obvious that improved stove design and good ventilation can reduce this problem. However, good ventilation that releases pollutants into the outside air can run foul of the law. In Alaska, during 2015/2016, the U.S. Environmental Protection Agency (EPA) was coming down on wood burning for heating in low population density areas because of the emission of fine size particles and gases through efficient chimneys to the atmosphere but at locations where there was little, if any, public health threat. Although the indoor air pollution hazard to health is not a problem in the United States, the EPA is reviewing their laws with respect to such conditions suffered by citizens in other countries. One temporary solution would be to pass legislation that requires good ventilation in homes where solid energy sources are used until homes are serviced by alternate energy sources. This, however, increases outdoor air pollution such as has been the case in Beijing and may not be a good solution depending upon meteorological history. Plans to reduce greatly or eliminate indoor air pollution are more easily proposed than immediately implemented. They include making low-cost (subsidized) fuel oil available for use in abodes with good ventilation and/or bring natural gas and electricity to where they are lacking. This option is possible with low costs loans from international or regional lending institutions or outright grants from rich countries and in some cases from non-governmental organizations (NGOs). The benefit to public health in some cities with high population zones and where incomes are low is costly but represents long-term investment that can save millions of lives globally over time as the proposed options are brought on line. An economic benefit is a healthy workforce that assures stable or increased productivity.

Indoor air pollution should not be a problem in a newly built city where building codes require that residences, commercial structures, and manufacturing/industrial facilities have efficient pollution control systems that capture pollutants before moving cleansed air out through efficient ventilation systems. One concern is if rural in-migration brings in populations that give rise to large shantytowns or slums where combustion of solid energy sources is not controlled. As defined in the prologue, a shantytown or slum is a densely populated (crowded), dirty, run down, squalid part of a city consisting of a large number of crudely built dwellings inhabited by poor people often lacking in social organization and underserved with the basic necessities of life. Governments can invest to redo shantytowns with proper housing and utilities or do as Nigeria in Abuja City where the government legislated and put into practice the razing of shantytowns within and abutting the city and resettling hundreds of thousands of displaced Nigerians in other locations.

3.2.1.3 Human Activity Caused Outdoor Air Pollution: Problem Solutions

Some sources of outdoor pollution that can engulf a population center for days or longer periods are related in part to poor siting. The principal sources are industrial zones that are located in the outskirts and upwind of an urban center and urban vehicular emissions. The siting of these zones that bring dangerous levels of outdoor air pollution to population centers failed to consider meteorological conditions that carry emitted fine size particles, and toxic chemicals and heavy metals as gases and aerosols, downwind to a city population. This often contributes to smog that puts citizens at risk for periods of a few days but can last longer. This happens mainly during winters, because of cold air inversions and stagnant weather fronts. As mentioned previously, fine size particles from desert sandstorms that can go on for days may contribute seasonally to the outdoor bad air problem. The focus on rehabilitation planning to improve citizens' public health and their quality of life has to focus on multiple solutions. A main one would be to prevent or greatly reduce emissions from manufacturing and industrial plants that discharge fine size particles and toxic gases and aerosols to the atmosphere by 90% or more. This includes coal fired electricity generating plants and cement factories among major polluters. Another would be reduction of emissions into the atmosphere from trucks, cars, buses, and motor scooters. Concentration build up, and interactions, of emissions from all sources as a result of lack of movement of weather fronts can create dangerous outdoor public health air pollution conditions. It can also affect economies in cities where visibility is so reduced as to stop vehicular traffic and bring air traffic to a halt for a period of time.

One solution to the fixed emitters is not economically viable when there are many of them. That would be, for example, to move manufacturing, industrial, and coal-fired electricity generating facilities downwind of population centers. Two others solutions to greatly reduce greatly fine size particles and toxin emissions are to install and use the best available pollutant capture and control systems in chimneys such as electrical precipitators to capture fine size particles and chemical scrubbers to capture harmful elements and compounds. A third option is to simply to close down facilities that opt not to either move or install pollution control equipment. The switch to natural gas from coal as an industrial energy source can reduce particle, heavy metal and other emissions that originate from coal combustion. This solution applies as well to well ventilated stoves that otherwise use "brown" coal for cooking and winter heating and emit fine size particulates and some metal toxins to the atmosphere in populated cities, especially in China. In many countries industries are switching to natural gas because it is cheaper than coal. Government officials have to measure the benefits (many immediate) against the losses (few in the long run), when opting to use development funds to bring natural gas and/or electricity to where it can replace coal as an energy source. In recent years, solar systems have brought electricity to small population centers in Africa and Asia but often these serve for lighting and small appliances but not enough for cooking and heating. Nuclear electricity generating facilities continue to be built and developed with fail safe systems being of prime interest. Long-term planning for economic sustainability to replace short-term economic development gain has governments using available funds for environment good as more opt for future sustainability.

The reduction of harmful emissions in urban centers from vehicles with gasoline (or diesel) internal combustion engines is possible in two or three ways. One would be the development of an efficient electricity-based urban transportation system (trackless trolleys, trams, subways) where a long-term investment can reduce gasoline engine vehicular use to a huge degree. Another is to capture gases before they are emitted from vehicles using catalytic converters but this adds expense to vehicles and their operation. This would be politically unpopular in less developed and developing countries (e.g., India). A third would be the permanent alternate day odd license plate-even license plate access to a populated urban center or even to ban unnecessary vehicular traffic in some city areas (e.g., Central London). Certainly, the future will dictate use of electric cars to stop the vehicular source of outdoor air pollution. All new cars will be electric in Germany by 2030. The Netherlands announced a ban on petrol and diesel cars by 2030 as did France and Britain by 2040. It will be opposed by oil companies without electricity generating subsidiaries but embraced by electricity utility interests.

New Delhi, India has an outdoor pollution problem that seasonally puts public health at risk. This is the burning of the straw and stubble residue after a rice harvest or a wheat harvest [10]. This action sends fine size particles and dense smoke over Delhi for days on end. Many citizens crowd into health clinics and hospitals with severe bronchial problems. The burning takes place in an adjacent state so that the Delhi government can not stop it. It then is the responsibility of the federal government to do so but it does not do so. Prime Minister Modi's government should be able to act in defense of the health of millions of Delhi citizens. This might be based on a no-burning dictum or better yet on education that shows rice and wheat farmers that plowing the crop residue back into the soil can improve the nutrient content. This would maintain crop production without the addition of as much fertilizer as they now use resulting in lower production costs.

3.2.2 Water Pollution: New City

The planning of a new city can effectively remove the threat from water pollution with a master plan that first designates sites for industrial parks at some distance and down waterway and aquifer flow from the new city. A second safeguard is to put in place legislation that requires manufacturing and industrial facilities and other potential polluting sources to treat effluents to remove high percentages of toxins (e.g., >90%) before their discharge and to monitor treatment use and efficiency. In addition, this requirement of treatment to remove toxins before issuing effluent into the environment preserves the viability of land designated for future urban expansion at the city outskirts including open space and farther out, soils for agricultural ventures. Clean waterways and aquifers will serve citizens of a new city.

3.2.3 Water Pollution: Established Urban Center

A rehabilitation/redevelopment plan to stop water pollution that causes public health stress in existing heavily populated urban centers is complex and more so with their growing populations. In some cases, polluting entities are located in the midst of urban centers in addition to polluters in city outskirts. In both scenarios the locations or inefficient treatment of polluted waters before effluent discharge can contaminate city surface and/or aquifer water supplies.

Water pollution is an existential problem for many highly populated cities in less developed and developing nations. This presents a more complex problem to deal with especially for growing populations in city limits but more so in shantytowns/slums that have built up skirting a city. A major issue is that there may be less water per capita to service an increasing number of people and also commercial, industrial, and agricultural needs. This is especially true for mega-cities (>10 million people) or other highly populated centers. If the effluent discharged from manufacturing/industrial facilities is tainted bacterially or with toxic inorganic and/or organic chemicals, or if toxins infiltrate through soils into aquifers, still less per capita useable water would be available from otherwise tappable water sources without passing through a water treatment facility. A major public health problem tied to water exists where there are large and dense concentrations of citizens without adequate access to sanitation facilities. The result is that sewage is discharged into open waterways that are used by poor citizens. This causes hundreds of millions of cases of diarrheal sicknesses annually. In some cases, epidemics such as cholera can take hold depending on the personnel and medical supplies capability to take control of such an infectious disease condition. The result is a loss of productivity and a shrinking per capita GDP as workers lose days wages because they are unable to function. Drainage from poorly sited waste disposal dumps into surface water sources or leakage that infiltrates aquifers carries contaminants that can sicken populations that use the tainted waters. The solutions to this problem is investment in sewer systems that collect and treat sewage, and in the siting of waste disposal dumps away from water sources. In addition, the dumps can be engineered with sloped impermeable linings to collect leachate from rain infiltration or surface water runoff in tanks and recover it for subsequent treatment. Again this is a costly undertaking during rehabilitation of existing unserviced areas of population centers. Nonetheless, it is worth the investment because of future economic payback by less of an economic draw on health facilities, and from sustainable and improved worker health and productivity (less lost days).

In addition to a population's health risk from the use of polluted water for drinking, cooking, and personal hygiene, people are at risk as well if they consume crops that absorb toxins grown in soils that build up (bioaccumulate) over time in body organs to concentrations that can cause health problems. This may be from natural concentrations of toxins in soils or from tainted irrigation water or contaminant bearing industrial effluents discharged onto soils. As already noted, the discharge of "clean" effluents into waterways that keep existing urban and rural populations healthy that may receive runoff from nearby industries can be achieved by the installation of the best available, most efficient equipment that removes (captures) toxins from effluents before release onto soils or into waterways. Such installations can be required by local or national government legislation. Legislation can be effective only if it has enforcement "teeth". This maybe via meaningful fines or closure of a facility if installed pollution reduction equipment is not used 24/7 during production operations. As with all "fixes", pollution control is costly but a timely investment that can be borne by industries planning for continued or improved productivity and payback in a reasonable timeframe.

3.3 Subsidence: A Worldwide Water Problem Tied to Human Activities and Flood Risk

Subsidence is a lowering of the land surface. As such it can damage buildings, disrupt infrastructure, enhance the effects of flooding, and in coastal areas extend the reach inland of high energy tropical storms and storm surges.

3.3.1 Causes and Effects of Subsidence

The principal causes of subsidence are the extraction of groundwater from underlying aquifers and the extraction of petroleum and natural gas from their subsurface reservoirs. These fluids impart strength to rocks by exerting buoyancy pressure. When extraction lowers the buoyancy pressure and recharge of fluid does not restore it, subsidence can take place. With respect to existing cities, especially heavily populated ones, rapid expansion and urbanization from population growth raises demands on water supplies. When groundwater is the main source, subsidence can take place if recharge does not match discharge. Recharge would naturally take place as rainwater or melting snow or ice seeps through soils into an unconfined aquifer. Recharge becomes greatly limited when open land is covered by residential and industrial structures and by roads and sidewalks. This leads to excessive runoff during heavy rains and exacerbates flooding that takes place in addition to major reductions in aquifer recharge.

Because it lowers a land surface, subsidence can subject population centers to more widespread flooding, longer lasting flooding, and higher inundation depths. In addition, subsidence damages structures and disrupts infrastructure from flood control elements, broken sidewalks and roads, undermined bridge supports, to ruptured water/sewage lines or reversals in them that causes backups of wastes. Also, aqueduct reversals that can further affect water delivery as was the case in California's Central Valley during 2014–2015. Together these cause major disruptions of water management controls. In coastal areas, subsidence coupled with sea level rise puts populations at still greater risk of injury or death from tidal forces and storm surges that coupled with torrential rain can cause widespread inland flooding.

Subsidence can also be the result of a natural process such as compaction of alluvium soil and can lower the surface 1 mm/year or more. It can also result from extraction of groundwater from peat soils that exposes them to oxidation and mass/volume loss. The weight or load from building construction can cause a compaction of underlying earth material and lowering of a land surface. The degree of compaction can vary at a building site (differential subsidence) and can be calculated by analysis of the bearing strength of soil and underlying rock. Failure to correctly determine the potential for subsidence/differential subsidence can result in problems for a structure. For example, the newly constructed Millenium Tower, a 58 story condominium in San Francisco, USA, has settled 16 in. (6+ cm) and tilted at the top 15 in. (6 cm) to the northwest as the result of differential subsidence. It continues to subside. Concrete piling were sunk into mud, sand, debris earth materials in the subsurface to 80 ft. (24+ m). The subsidence would not likely have taken place if the pilings, at additional construction costs, were anchored in bedrock at 200 ft (61 m) in the subsurface. If the tilt increases, high speed elevators may stop functioning. It is surprising that building inspectors did not require anchoring in bedrock given the mass of the structure and the fact that the San Francisco region is prone to earthquakes. Earthquake resistance mechanisms were part of the construction plan following the requirements of the existing building code. Whether they will stabilize the slightly titled structure during a major earthquake is something to consider for those who own Millennium apartment homes or those with desires to buy one.

3.3.2 Examples of Urban Subsidence

Subsidence is an ongoing worldwide problem but is most threatening to major population centers where urbanization with growing populations (natural growth, influx from rural areas, immigration) causes municipalities and citizens to pump more water from aquifers than is being recharged. The loss of previously cited buoyancy pressure allows compaction of aquifer sediments and the slow subsidence of the land. In a few cases, cities (e.g., Tokyo, Venice, Beijing) have found, or created by river diversion, alternate sources of water so that subsidence has been greatly slowed or arrested. But this is not the case in many major urban centers with populations in the millions and mega-cities with populations greater than 10 million people. These include Jakarta, Bangkok, Ho Chi Minh City, Manila, Shanghai, and Mexico City.

The amount of annual subsidence can vary greatly within the area of a population center and recorded maximum cumulative subsidence shows great differences between affected areas. This depends in part on the thickness of aquifer sediments, whether one or multiple aquifers are being pumped, and on the type of aquifer sediment that underlies a land area such as coastal (deltaic), river, or filled lake sediments. Tokyo was extracting water from multiple aquifers (coastal sediments) and differential subsidence was measured in the Tokyo plain. During the 1960s, the government dictated a stringent lessening of water extraction by withdrawal management, groundwater level monitoring, and sought alternate sources of water including promotion of rain water infiltration [11, 12]. Today, recharge of aquifer water is in equilibrium with discharge so that there is zero subsidence (stabilization) vs. The total cumulative maximum subsidence registered at one locale of more than 4 m (~13 ft).

3.3.2.1 Ho Chi Minh City (Formerly Saigon)

Ho Chi Minh City has many of its 7.5 million citizens living at elevations of 1–5 m (3 m average [~10 ft]) above sea level. Urbanization from the influx of people from rural areas has raised the demand for water pumped from aquifers in subsurface alluvial sediments. This, together with loading from construction and from 50% of land formerly a groundwater recharge area now covered with concrete is resulting in subsidence of 1–2 cm annually (<1 in.) [13]. Groundwater is being extracted at 1 million m^3/day whereas the recharge is at 200,000 m^3/day. The use of permeable pavement in redevelopment projects as may be planned would reduce flooding and allow a stronger recharge of aquifers. Differential settling/subsidence is highest where there is heavy urbanization. Couple the subsidence of 1–2 cm (<1 in.) annually with a sea level rise of 3.3 mm annually and the submergence of some coastal regions is only a few generations in the future unless subsidence is greatly reduced. The existing situation in Ho Chi Minh City sees high tide flooding of many important roads and puts citizens at risk during storm surges at coastal population centers. Economically, repairs of infrastructure and structures as the result of subsidence costs Viet Nam municipalities and individuals millions of dollars annually. There is the option of building sea walls in coastal areas most susceptible to flooding and damage from storm driven seawater surges but this is extremely costly and out of the reach of many threatened coastal population centers without financial assistance from international banking organizations and grants from financially strong nations [14].

3.3.2.2 Bangkok

Bangkok, a coastal mega-city with more than 14 million people, has subsidence problems that are similar to those described above for Ho Chi Minh City. Topographically, the mean elevation above sea level is 1.5 m (~ 5 ft) with the elevation decreasing to 1 meter moving south towards the coast. As in most urban subsidence locales, the main cause is extraction of water from underlying aquifers from municipality registered wells as well as unregistered ones for private use. Past loading of the underlying soft, weak clay soil with the weight of buildings (700 with 20 or more floors and 4000 with 5–20 floors) and infrastructure adds to Bangkok's subsidence problem [15, 16]. The use of floating foundations could have mitigated this additional cause of subsidence. During the 4 decades from 1976, subsidence expanded into newly urbanized areas with increased demand for water. The cumulative subsidence is more than 1 m and the present average rate of subsidence is 1–2 cm (<1 in.) annually with the higher value where there is a higher pumping of water, often for manufacturing/industrial use rather than domestic/commercial use. Flooding increases from high stream flow during rain storms because of loss of open land and canals that carried water to the ocean and from concrete covered sidewalks and roads and residential and industrial structures where water would formerly infiltrate the subsurface but now runs off into subsided areas. There is more flooding along the coast with high tides. At an average rate of subsidence for Bangkok of 1.5 cm/year (<1 in.) more than 1 ft of elevation (>30 cm) will be lost every 20 years. This, plus 3.3 mm annual predicted sea level rise, puts Bangkok's coastal and inland populations, infrastructure, and property at risk from sea water encroachment and from tropical storms and storm surges and flooding that may accompany them.

One solution to reduce flooding and coastal hazards would be to relocate Bangkok inland but clearly, this is not feasible economically or socially. Another proposal to protect Bangkok is to build a 100 km (62 mi) long sea wall across the Gulf of Thailand to keep out rising sea level. At a projected cost of an estimated US$14.3 billion this is feasible but an economic burden for the country and may not be a lasting solution. The obvious first step to reduce subsidence is to greatly reduce water extraction from underlying aquifers by registered wells and especially from unregistered wells. This can be complemented by putting in wells that move surface waters under appropriate pressure (hydrogeology decision) to arrest subsidence and possibly induce some recharge and rebound. This should be coupled with the construction of an infrastructure that moves water from rivers to water treatment plants and distribution networks that transport clean water to citizenry for domestic and commercial use, and as needed, to industrial/manufacturing facilities.

3.3.2.3 Mexico City

Mexico City is built over a lake (Lake Texcoco) that was drained in the early 1600s by the Spanish conquistadores [17, 18]. This mega-city (>21 million people) draws 70% of its water supply from aquifers beneath the city in clay and readily compacted sand and gravel sediments. Twenty-three percent of the water supply is pumped from outside the city area and 7% is recycled. The water problem is further exacerbated by the water ministry estimate that ~40% of the water pumped to the city is lost to leaks. During the last century, subsidence was ~9 m (~30 ft). Differential settling/subsidence leaves some structures dangerously tilted (leaning) and other not. Structures built on pilings down to bedrock appear to grow as the surrounding ground settles beneath them. One above ground Metro line that was level when built now appears as low waves as a result of different subsidence rates along the metro route. The Palacio de Bellas Artes (built 1934) has subsided so much that its former entrance is below ground level and the Palacio is entered by the first floor. The Monumento a la Independencia (built 1910) originally had 9 steps up to its entrance but now has 23 stairs up to its entrance as the land around it subsides. Many affected structures are World Heritage sites. Areas in the city continue to subside at an average rate of 30 to 40 cm (12–16 in.) annually as determined by remote sensing techniques [19, 20]. Restoration projects for historically important structures are active. Slow down in the rates of subsidence can be achieved only of there is a meaningful reduction in the rate of water extraction. Groundwater and surface water management teams will be pressed to serve another few million people by 2050. This requires that investment be make to seal the leaks in the system that carries water to the city. It means that an infrastructure has to be put in place that collects Mexico City's abundant rainfall and move it to treatment plants for cleansing and subsequent distribution to users. Investments have to be made in recycling operations so that "dirty" domestic waste water is collected, put through treatment plants and redistributed to users. Mexico has an economy that can contribute to this plan to alleviate the capital city's clean water problem. This or a similar management scheme should make coping with water needs of the existing population and that added to by population growth a solvable problem. A solution that has been posited is to move the capital city to a physically and environmentally sustainable location.

3.3.2.4 Houston, Texas

Houston, Texas has a population of ~2.5 million people and is projected to double this number by 2050. Metropolitan Houston has a population of 6.5 million people. The Houston area (Harris County to Galveston County) suffered from subsidence of up to 2 ft (61 cm) from oil and gas extraction from the Goose Creek Field initially from 1903–1918 with damage to homes and roads. There was an additional ~10 ft (3 m) subsidence in some locales from over usage of water extraction from aquifers. The need to cope with subsidence was emphasized to communities following the destruction wreaked in the region from Hurricane Carla in 1961. The subsidence contributed to the frequency and severity of flooding in the area as well as fracturing of roads, rupturing of important parts of industrial operations (Port of Galveston, oil refineries, pipes), and problems with drainage systems such as fractured sewer lines and canals, and loss of water/gas/ electric utilities, all of which were very costly to repair and put back into operation. Most of these problems affect growing urban centers that suffer subsidence mainly from extraction of water from aquifers to service urbanization with its influx of population and manufacturing/industrial development. In 1975, legislated withdrawals of groundwater were put in place for Harris and Galveston counties and by 1985 there was an ~50% drop in groundwater extraction. Because of differential subsidence in the area, withdrawal allowances vary with locale. During the 1990s, subsidence was essentially stopped because of a great reduction (80–90%) in the use of ground water replaced by the use of surface (river and lake) waters. Part of the reduction of groundwater extraction was the result of a disincentive fee to users of $3.00/1000 gallons of water [21–23].

At the end of August, 2017, hurricane Harvey impacted Houston and many surrounding communities with high velocity damaging winds and torrential rain. During 4 days, the hurricane moved little and generally dumped from about 2 feet to more than 3.5 ft (~1 m) of rain on the area as the storm continually sucked up moisture from the warm Gulf of Mexico waters. The ground became saturated quickly and flood control could not contain the runoff waters resulting in the damaging, destructive flooding. Subsided areas, such as described in the previous paragraph, and flat terrain in the region, including Houston, flooded to record breaking levels reaching a 16 ft. (4.8 m) height in one area, killing 65 people, injuring many others physically and traumatizing thousands while severely damaging or destroying more than 40,000 homes and commercial structures, as well as battering infrastructure and industrial operations especially oil refineries, and port facilities. The area contains 41 US Environmental Protection Agency (USEPA) Superfund sites many of which released toxic matter into the environment during the flooding. The cost to restore physical normality to Houston and the surrounding population centers will likely be close to $200 billion, much from owners uninsured against flooding. In comparison, the cost of Hurricane Katrina that ravaged New Orleans in August, 2005, was $160 billion in 2017 dollars. People will have to decide whether to repair homes if possible, whereas others will have to decide whether to rebuild. Flood control will have to be greatly improved to maximize flood protection for people and property from future storms. Zoning will have to be legislated that prevents habitation where risk of flooding is high such as on a 100 year flood plain. Building codes will have to be modified to reflect needed improvements to structures in light of high winds and recurrence of flooding when torrential rain storms blanket the region. The USEPA will have to invest in solving public health and environmental problems caused by the release of toxic matter from damaged Superfund sites.

3.3.2.5 Jakarta

Jakarta, Indonesia is another coastal mega-city with more than 10 million inhabitants and with agglomerated populations has more than 28 million people. The city is situated in a lowland (0–2°slope) underlain by deltaic deposits. It has a multitude of environmental problems with subsidence the major one that puts the mega-city's future existence at risk. As in most urban centers with varying degrees of subsidence, the lowering of the land surface is the result of over usage of aquifer waters because discharge from aquifers exceeds recharge. This is the result of rapid urbanization of agricultural and other open land where recharge once took place plus the need to deliver water to growing populations and for expanding industrial development. The loss of strength lent to aquifer sediments by buoyancy pressure resulted in a compaction of sediment and subsidence of Jakarta's land surface. Coastal development was favored with rich and poor residential areas and fishing ports, all with their infrastructure. Today they are sinking as is a 40 year old sea wall built to protect the coastal development, some areas of which are close to 2 m (6.5 ft) below mean sea level and subject to regular tidal and storm surge flooding. Generally, where withdrawal of water from an aquifer is greatest is where rate of subsidence is more pronounced. Others factors that add to subsidence in Jakarta are heavy loading by buildings and infrastructure, natural settling (compaction) of the underlying soil and sedimentary sequences, and possibly tectonic activity. Thirteen rivers and two canals flow across the city from the southern area that has a maximum elevation of 50 m (164 ft) above mean sea level. River beds are sinking as well so that for about 1.5 km (~1 mi) some of the waste carrying (sewage included) river courses were reversed and they could not flow upslope to empty into the Java Sea. They had to have their channels routed into canals that emptied into the sea. Between 1974 and 2010 the average rate of subsidence varied from 3–10 cm (>1–4 in.) annually. The damage done as a result of subsidence mirrors that in other population centers that suffer or have suffered from subsidence. These are generally more pronounced in Jakarta and include increased river and coastal flooding with extended reach inland with greater frequency and longer duration, sea water encroachment inland (abetted by sea level rise), rupture of infrastructure including roads, bridges, drainage networks, and of utility delivery systems [24, 25].

Three options to deal with subsidence and other problems that put large segments of Jakarta's high density population at risk have been proposed. One is to move the city to an elevated area southeast of the city (50 m [164 ft] above mean sea level) or to another island. An estimated cost of US$220 billion makes such a move economically improbable. A second option is to abandon the old city district of north Jakarta and concentrate resources on preserving the rest of the city. Because of development in the north, this is not favored. The third option is to reenforce coastal defenses by first rehabilitating the canal flood drainage system and by building a Giant Sea Wall (systems of coastal and estuary levees) at an estimated cost of US$40 billion [26, 27]. This would require financial assistance via grants from economically strong nations or long-term low cost loans from international lending agencies such as the World Bank or the Asian Infrastructure Development Bank. This is a longer term option but not a permanent solution unless withdrawal of aquifer water is greatly reduced by enforcement of existing regulations, constructions of river water collection and treatment plants, and piped distribution of clean water to populations. Lacking this now, subsidence will continue and together with other threatening factors cited above and future climate change puts the continued existence of Jakarta as we know it today in question.

3.3.2.6 Netherlands

In addition to the causes of subsidence repeatedly cited above, another cause of special concern in the Netherlands is the degradation of peat soils. This is also a worldwide problem. About 9% of the Netherlands, mainly in the west, is underlain by 20–40 cm (8–16 in.) of clay overlying several meters of peat. To keep the land suitable for agriculture the peat has to be seasonally drained. This lowers the water table and exposes the peat to oxidation that causes a loss of volume, compression of the peat beds and subsidence of the land surface. The oxidation also releases CO_2 to the atmosphere [28]. This has become an economic burden because of the need to immediately repair cracks in structures and in roads/sidewalks and ruptures of sewers and underground utilities (gas, water, electricity, communications) at a high cost borne by the citizenry. The agricultural areas have deep ditches at 6 m (~20 ft) spacing that are filled with water to raise a water table between them or drained of water to lower the water table. The peat subsidence has been slowed by inletting water to replace the drained groundwater after a growing season thus limiting oxidation and resulting peat compaction and out letting water for the new growing season. This method has been modified using submerged drain infiltration that is more efficient and keeps water levels between 40 cm (16 in.) in the winter and 60 cm (24 in.) in the summer [29]. Without the submerged drains the water levels were greater than 80 cm (32 in.) in the summer exposing more peat to oxidation while during the winter the water level is closer to the surface at 30 cm (12 in.) or less deep thus limiting peat exposure to oxidation. By reducing the peat exposure to seasonal oxidation, subsidence has been reduced to between 2–25 mm (~1 in. maximum) annually [30].

3.4 Wildfires

Wildfires can be caused by natural forces such as lightening, spontaneous combustion, wind/storm downed power line arcing, and by volcanic eruptions. These account for 10–20% of wildfires. Eighty to 90% of wildfires are mainly the result of arson abetted (10%) by human carelessness. The latter includes unattended camp fires, unexpired and tossed tobacco ash, poorly controlled burning of debris, sparks from train wheels and operation of machines, land conversion burning (e.g., in Brazil). and in some cases prescribed burning to destroy vegetation fuel when there is a wildfire and a sudden change in climatic conditions. The intensity and spread of a wildfire depends on fuel (e.g., dry vegetation), climate (temperature, relative humidity), wind velocity and dominant wind direction, and topography (fires are drawn upslope). Planning for a new city has to take into account the potential of wildfires based on an areas history and build with mitigation methods in place if a proposed city site has been impacted by wildfires.

Wildfires can sometimes affect major cities with high populations often through the ash and smoke they generate carried downwind of a fire. This threatens public health through inhalation of fine size particulates (<2.5 μ) by citizens with chronic bronchial or other conditions (e.g., heart disease, emphysema, asthma). Five wildfires (bush fires) around Sydney, Australia during October, 2013 sent a plume of ash and smoke that blanketed the city of almost 5 million people and caused a respiratory alert to be sounded by public health authorities. The plume was so dense that it showed up on radar sweeps. The cause of one fire was a downed power line but others were the result of arson. Property and businesses lost to wildfires affect not only people but also economies. In 2016, wildfires in Chile raced through vineyards of globally well known wineries causing losses in production and wines for future export. In 2017, wildfires in northern and southern California, USA have killed at least 43 people and burned down thousands of homes and businesses and roared through a large area of vineyards of well known wineries in the north and residential/commercial areas in the south. The property and economic losses will be vast from burnt structures and vines, and lost tourism. There is really no protection against the start of a fire but if spotted immediately from ranger observation towers, fire fighters can be dispatched and a nascent wildfire may be brought under control before spreading. Very often the start of a potential wildfire is not spotted before it becomes an uncontrolled wind driven and fast moving hazard that firefighters, at great risk to themselves, work to bring under control and extinguish. Neighborhoods that could be at wildfire risk can work to thin or remove (especially dry) vegetation that could be fuel for an advancing fire. New construction in these areas should use fire resistant materials.

3.5 Sabotage and Terrorism in Today's World

Sabotage has been responsible for industrial hazards carried out by disgruntled employees. Depending on an industry's location, released toxins can sicken, injure, and kill citizens (e.g., Bhopal, India). In addition to sabotage, industrial facilities have released toxins because of operator error or natural hazard impact (e.g., Chernobyl and Fukushima, respectively). Industrial facilities are difficult to protect from sabotage and from human errors in operations or siting but dangers to people can be mitigated if industries are distant and downwind from population centers and away from locations that have suffered from natural hazards. Blowing up of oil rigs and/or pipelines are acts of terrorism carried out by religious fanatics, repressed people, irrational activists, and retreating armies. This can cause massive, long lasting fires with heavy smoke that endangers people and ecosystems/environments that may be downwind. Terrorism relies on indiscriminate acts of violence designed to create terror and fear worldwide (e.g., in the United States, England, France, Germany, Lebanon, Syria, Russia, Indonesia, the Philippines) but instead reinforces resiliency in national populations. It can be stopped only when wars end or when terrorists realize that their efforts very rarely achieve the political, idealogical, or religious aim they purport to bring about. It is difficult to predict who might be planning terrorist actions to create and maintain fear in a population so as to be able to prevent them, but many attacks are thwarted by intelligence services or informant warnings.

References

1. Siegel, F. R. (2016). *Mitigation of dangers from natural and anthropogenic hazards: Prediction, prevention, and preparedness*. New York: Springer, 127p.
2. California Building Standards Commission. (2013). *California existing building code* (752p, Vols. 1 and 2). State of California: Sacramento.
3. Teshigawara, M. (2012). Appendix A: Outline of earthquake provisions in the Japanese building code. In *In Preliminary reconnaissance report of the 2011 Tohoku-Chica Taiheigo-Oko earthquake, geotechnical, geological, and earthquake engineering* (Vol. 23, pp. 421–446). Tokyo: Springer.
4. Minoura, K., Imamura, F., Sugawara, D., Kono, Y., & Iwashita, T. (2001). The 869 Kogan tsunami deposit and recurrence interval of large scale tsunamis on the Pacific coast of Northeast Japan. *Journal Natural Disaster Science, 23*, 83–88.
5. Siegel, F. R. (1996). *Natural and Anthropogenic Hazards in Development Planning*. San Diego: R.G. Landes Company, Austin, and Academic Press, 300p.
6. World Health Organization. (2014). News release: 7 million premature deaths annually linked to air pollution. Online. www.who.int/mediacentre/news/releases/2014/air-pollution/en/
7. Lancet. (2017, October 19). The Lancet Commission on Pollution and Health. 51p, 418 references. https://doi.org/10.1016/S0140-6736(17)32345-0.
8. IHME. (2017). *State of global air/2017. A special report*. Boston: Health Effects Institute, 15p.
9. Nkrumah, J. D., Okine, E. K., Mathison, G. W., Schmid, K., Li, C., Basarab, J. A., Price, M. A., Wang, Z., & Moore, S. S. (2006). Relation of feedlot feed efficiency, performance, and feeding behavior with metabolic rate, methane production, and energy partitioning in beef cattle. *Journal of Animal Science, 84*, 145–153.
10. Damodaran, H. (2017, November 10). Delhi air pollution: A (crop) burning issue, and the way out. *The Indian Express*.
11. Yamamoto, S. (1984). Case history No. 9–4, Tokyo, Japan. In J. F. Poland (Ed.), *Guidebook to studies of land subsidence due to ground-water withdrawal*. Chapter 9-4, (pp. 175–184). wwwrcmnl.wr.usgs.gov/rgws/Unesco/PDF-Chapters/Chapter9-4.pdf
12. Nakajima, H., Kaneko, H., & Minoru, T. (2012). The management of land subsidence and groundwater conservation in Tokyo. *Journal of Groundwater Hydrology, 52*, 35–47.
13. Nguyen, Q. T. (2016). The main causes of land subsidence in Ho Chi Minh City. *Procedia Engineering, 142*, 334–341. Elsevier Open Access. https://doi.org/10.1016/j.proeng.2016.01.058.
14. Tran Ngoc, T. D., Perset, M., Strady, E., Phan, T. S. H., Vachaud, G., Quertamp, F., & Gratiot, N. (2016). Ho Chi Minh City growing with water-related challenges. In *Water megacities and global change*, 27p. eaumega.org/wp-content/uploads/2016/05/HCMC-MonographyEN.df
15. Phien-Wei, N., Giao, P. H., & Nutalaya, P. (2006). Land subsidence in Bangkok, Thailand. *Engineering Geology, 82*, 187–201.
16. Osathanon, P. (2015, July 23). Action required to stop sinking of the capital. *The Nation*, Thailand Portal.
17. National Research Council. (1995). Chapter 3. Description of the Mexico City aquifer and its exploitation. In *Mexico City's water supply: Improving the outlook for sustainability* (pp. 8–18). Washington, DC: National Academy of Sciences.
18. Rudolph, M. (2001, July). Sinking of a titanic city. *Geotimes*. Washington, DC.
19. Chaussard, E., Widowinski, S., Cabral-Cano, E., & Amelung, F. (2014). Subsidence in Central Mexico detected by ALOS InSar. *Remote Sensing of the Environment, 140*, 94–106.
20. Sowler, A., Bin Che Amat, M., Cigna, F., Marsh, S., Athab, A., & Alshamman, L. (2016). Mexico City land subsidence in 2014–2015 with Sentinel-1 IW TOPS: Results using the Intermittent SBAS (ISBAS) technique. *International Journal of Applied Earth Observation and Geoinformation, 62*, 230–242.

21. Coplin, L. S., & Galloway, D. (1999). Houston-Galveston, Texas: Managing coastal subsidence. In D. Galloway, D. R. Jones, & F. Ingebritsen (Eds.), *Land subsidence in the United States* (pp. 35–48). Washington, DC: USGS Circ. 1182.
22. Neighbors, R. J. (2003). Sunsidence in the greater Houston area – past, present and future. In *Proceedings, CIGMAT Conf., Part 1, Presentations, I-1-I-3*. Washington, DC, cigmat.cive.uh.edu/conference/presentation/2003
23. Galloway, D. L., Coplin, L. S., & Ingebritsen, S. E. (2003). Chapter 12. Effects of land subsidence in the Greater Houston area. In D. C. Agthe, R. B. Billings, & N. Buras (Eds.), *Managing urban water supply* (Vol. 46, pp. 187–203). Water Science Technology Library.
24. Abidin, H. Z., Andreas, H., Gumilar, I., Fukuda, Y., Pohan, Y. E., & Deguchi, T. (2011). Land subsidence in Jakarta (Indonesia) and its relation with urban development. *Natural Hazards, 59*, 1–19.
25. Abidin, H., Andreas, H., Gumilar, J., & Brinkman, J. J. (2015). Study on the risks and impacts of land subsidence in Jakarta. In *Proc. Int'l Assn. of Hydrological Science* (Vol. 372, pp 115–120). Open Access. https://doi.org/10.5194/pihas-372-115-2015.
26. Republik Indonesia. (2014). Master plan, final, National Capital Integrated Coastal Development Project. 131p. On Line. www.en.ncicd.com/wp-content/uploads/2015/02/MP-final-NCICD-dec-low--res.pdf
27. TEMPO.CO. (2017, October 13). NCICD project expedited to prevent land subsidence.
28. Van der Akker, J., Hendriks, R., Hoving, I., & Pleijter, M. (2015). *Subsidence and CO2 emissions of Dutch peat soils in agricultural use and submerged drain infiltration to diminish peat oxidation*. Online at edepot.wur.nl/408687
29. Recare. (2018). Case study – Organic matter loss in organic soil experiment. Veenweidegebied, Netherlands. Online. www.recare-hub.eu/case-studies/veenweidegebied-the-netherlands
30. Erkens, G., van der Meulen, M., & Middlecoop, H. (2016). Double trouble: Subsidence and CO_2 respiration due to 1000 years of Dutch coastal peatlands cultivation. *Hydrogeology Journal, 24*, 551–568.

Chapter 4
Strategic Planning for Urban Centers: Present and Future

4.1 Planning a New City or Redeveloping a City: Needs and Spacing

Whether new cities will be built to accommodate increase populations that could no longer be served by existing city infrastructure and services is in question. The accommodations include clean water, sufficient food, acceptable housing, access to adequate sanitation, garbage/waste collection and secure waste disposal, transportation, personal security, and others. As noted in Chap. 1, during the latter half of the twentieth century three new major cities were purpose built: Brasilia, Brazil, from 1956–1960, Abuja, Nigeria, from 1982–1987, and Astana, Kazakhstan, 1994–1998. The planning for each was careful and precise according previously cited norms and spacing of components that made these urban centers sustainable as well functioning metropolises mindful of the well being of its population. Egypt planned a new city 45 km (28 mi) east of Cairo with construction to start in 2017 to serve 5 million Egyptians and thus relieve Cairene population pressure. The plan includes 1.1 million homes in 21 residential districts and 25 dedicated districts to house the 5 million inhabitants, creation of more than one million jobs, 2000 schools and colleges, 663 health care facilities, and 1250 mosques and churches, all in 5–7 years. However, the cost could not be borne by the country in 2017 and plans to start construction for the new city have been put on hold. China is also planning to build a new city, Xiongan, from scratch 62 mi (100 km) southwest of Beijing to reduce the population pressure (and air pollution) stressing China's capital city. The China Geological Survey has studied the area and found it good geologically and environmentally for surface and underground development as a new economic zone. The project received $19 billion in 2017 to start planning a "green" city to house 2.5–5 million people, accommodate some government offices, and receive industries that move from the Beijing area with the hope of reducing air pollution there. The project is expected to take 15 years to complete at a cost of more than US$290 billion [1, 2].

F. R. Siegel, *Cities and Mega-Cities*, SpringerBriefs in Geography,
https://doi.org/10.1007/978-3-319-93166-1_4

A combination of redevelopment of an old city and purpose building a new city is a continuing project for Astana, Kazakhstan. Astana is one of the oldest cities in the Euro-Asian region dating from 1830 when it was called Akmoly. The original name was changed four times before becoming Modern Astana when it was designated the capital of Kazakhstan in 1997. In 1998, a commission set forth a plan for redeveloping and preserving the old city and constructing Astana New City that would be the seat of government. This undertaking was mainly self financed. But foreign investment added US$1.3 billion for new buildings, roads, and a transit system for the Modern City. No cumulative real cost has been given. The population in 1999 was 281,000 but with new city construction and old city redevelopment with growing employment opportunities, the population in was 835,000 in the modern city in 2014 and grew rapidly to more than 1.3 million people in 2017. As has been the case in other cities with rapid population growth, Astana experiences air and water pollution. The air pollution is attributed to emissions from increased number of cars, plus buses and trucks that do not meet standards required, for example, in the European Union. Public health officials believe that this is the cause of increased cases of asthma and allergies they have treated. Untreated waste water (50%) and treated waste water that does not meet EU standards is believed to be the cause of acute intestinal infection in Astana. Ground water and water drawn from the Islim River is mineralized and contains zinc, iron, and sulfates pollution that officials believe could cause kidney and bladder diseases. The government is instituting more environmental control to mitigate or eliminate these public health pollution problems [3].

4.2 Spacing

In planning for a new city or for rehabilitation of an existing one that will be able to best serve the needs of existing or growing populations, an important factor that directs its development is spacing. Planners for Brasilia, Abuja, and Astana were well aware of this and the cities were constructed accordingly. The constraints of geology, geography, and locations of natural resources can determine the shape of a planned city from a center outward (e.g., as a circle or oval, with a square or rectangular shape, with an oblong shape) with well-planned spacing. Ideally, a center city and neighborhoods growing out from it would have a spacing that would seamlessly integrate residential (e.g., mixed housing), with local scale retail (e.g., food stores, drug stores, medical/dental offices, banks, electronics stores, clothing stores), with government institutions (e.g., schools, health clinics, municipal offices), and with cultural sites (e.g., museums, theaters, places of worship, parks). Then there would be larger scale retail (e.g., large supermarkets, major department stores, movie theaters, but no big-box stores [e.g., Costco, in the USA, Walmart globally]). The spacing would then zone to light manufacturing and light industry downwind and a distance way from residences and businesses but with easy public transportation to

Table 4.1 Arbitrary classification of population centers by numbers of citizens

<5000 = village/small town
5000 to 50,000 = small city
50,000 to 500,000 = medium to large city
500,000 to 1,000,000 = large city to high population city
1,000,000 to 5,000,000 = high population city to higher population city
5,000,000 to 10,000,000 = higher population city
>10,000,000 = mega-city

them. Lastly, heavy industries and landfills would be located down wind and well away (distanced) from populations' homes and businesses [3]. The scale of a city's components must fit for their effective use of space for the good of the citizenry.

Thus, we have a mixed-use blend of a city's fabric that has physical connections. A municipal master plan would reserve sites for new designated neighborhoods with their desirable component spacing to serve added population. In the case of outward (vs. vertical) expansion of urbanization, productive agricultural terrain at a city's outskirts should not be encroached upon but rather be preserved, thus maintaining a degree of food security.

4.3 Size: A Consideration in Planning Rehabbing Cities for Citizen Well-Being

The complexity of redeveloping existing cities to save people, property, and infrastructure from natural and man-made hazards that put citizens at risk of death, injury, or sickness is generally determined by the size of a city's population and its physical location. The complexity grows with the need to solve inequities as well in servicing basic needs, education, and other factors that affect the quality of lives of poorly served or unserviced citizens whether in city limits or in peripheral shantytowns. An arbitrary size classification is given in Table 4.1 and can be grouped into a lesser number of classes depending on who may be assessing local, regional, or national possibilities for attacking problems in urban centers. Here we must reemphasize the need to address projected continual growth of city populations to at least 2050. This is a little more than a generation in the future, and is most important for city planning in Asia, Africa, and to some degree Latin America. We must add to this the problems that have to be considered that might affect a municipality's ability to sustain growing city populations because of progressive changes from global warming and climate change such as in precipitation patterns that can affect water availability, and extended heat waves and drought that can affect health and food security.

4.4 Idealism and Major Cities/Mega-Cities Reality Problems: Planning and Coping

The focus of what **can be done** to improve the public health status and quality of life for those populations living in or moving into shantytown/slum conditions in major cities far overrides **what should be done** but is not feasible because of economic constraints. These may be satellite populations to mega-cities or cities with high (Table 4.1) and dense populations that are not serviced to the degree as are "developed" center city inhabitants. Nonetheless, they continue to attract rural in-migrants and immigrants seeking a better life for them and hence their families.

A problem faced by teams planning changes for existing urban centers that will ultimately benefit underserved populations is the continued population growth in a city's outskirt informal settlements. In many cases, the development is driven by economic growth that requires them to attract new commercial and industrial investment that promises a good return and more employment opportunities [4]. This leads to population increase as people migrate to fill jobs. This would liken to one step forward reaching an underserved population and one or two steps backwards as shantytown/slum populations grow at a faster rate than redevelopment. According to the United Nations, the shantytown/slum population worldwide grew from 689 million people in 1990 to 792 million in 2000 to 881 million people in 2014 as the global population grew from 6.1 billion to 7.2 billion people [5]. **Thus we come to a moral/ethical dilemma. Can a government or should a government pause the flow of in-migration and migration for a period of time until essential municipal services are available to citizens?** Are governments able to do this in societies, be they democratic, autocratic, or socialist (communist), in Asia and Africa where lack of services that abet public health and an improved quality of life are lacking for large segments of populations? If not, people will continue streaming to cities where they believe that there are employment opportunities, accessible health clinics, and education open to their children thus fueling unsustainable living conditions. China is solving the problem of overpopulation in Metropolitan Beijing with a plan to bring its population down from 24.9 million people in 2017 to a 'stable' 23 million in 2020. This is being done by "encouraging" migrant workers to leave. These workers who were already denied access to healthcare and free public education had their homes water, heating, and electricity cut off in mid-winter. Citizens were evicted from their homes that were demolished stating they were unsafe to live in. In January, 2017 the city authorized a project to raze more than 430 million ft^2 (~40 million m^2) of illegal buildings. Migrant workers in Metropolitan Beijing in 2017 numbered almost 8 million or about one-third of the city population [6]. A question Beijing citizens are asking is "Who will takeover the in-migrant citizens jobs of so many (almost 2 million) being forced out of the city by having their homes razed?"

4.4.1 The Essentials That Sustain Populations

If the flow of rural citizens into city outskirts and growth of shantytown conditions can be arrested, redevelopment planning that will improve living conditions for future increasing populations can be put in place in an orderly manner. New rehabilitated neighborhoods that replace shantytowns will serve all urban citizenry. Nonetheless, there are limitations to growth that involve the essentials to life that planning can not easily overcome. As stated in previous chapters, this is the availability of renewable surface and/or aquifer water sources to serve the maximum number of people that populate a city now and that may populate it in the future.

The renewable factor applies to both surface and subsurface water sources. For rivers, streams, and lakes, it requires that precipitation, rainfall and snowmelt be fairly constant seasonally to deliver renewable runoff. This may be a problem in assessing water availability for the future if global warming/climate change results in less precipitation and/or increased drought in some regions thus reducing their per capita water supply. A planning team has to consider this aspect by reviewing the most recent Intergovernmental Panel on Climate Change reports (IPCC.ch).

As emphasized earlier in the text, renewability requires that water discharged from an aquifer is matched by recharge, likely on a seasonal basis. If this is not the case, abetted by surreptitious citizen activity using unregistered wells to 'steal' water so that discharge continually exceeds recharge, an aquifer can be squeezed to the point that a diminished water source cannot supply this essential to life commodity to populations. Also, if compacting aquifer sediments reduces porosity and permeability, an otherwise rechargeable aquifer volume may be lost as an important water source in the future. As with surface waters, a planning team has to consider a future climate change that could reduce the amount of rainfall and snowmelt that serves to recharge unconfined aquifers. **It is essential for any redevelopment plan that the recharge areas for an aquifer not be covered with impermeable surfaces whether residences, sidewalks, roads all of which loses recharge waters to runoff and affects what one would hope is an equilibrium between discharge and recharge. The answer to this problem is to use pervious (porous, permeable) cement for surfaces overlying recharge areas.** In addition, we have reviewed the subsidence problem that puts many cities at risk of flooding from tropical storms and storm surges abetted by rising sea level (Sect. 3.2).

Treatment plants will definitely be necessary to clean available surface or subsurface waters or recycled domestic dirty waters. They would deliver the cleansed water through piped networks to residential areas and then, as needed, to commercial, manufacturing, and industrial facilities, and to agricultural projects. Water is the number one priority that must be considered for the present and for changes in the future as global warming/climate change affect water supplies.

The second major priority is food security without which there can be malnutrition and famine that can ravage a population. Planning for existing and future climate changes that shift agricultural zones and fisheries that could result in possible shortfalls in food production is essential if the lack of nutritious food for expanding populations, especially in urban centers, is to be avoided.

Once the existing and future carrying capacity of a population is assured for peoples' basic life sustaining needs, other factors that go into rehabilitation/redevelopment planning can be addressed. To repeat, these include public health security (e.g., access to adequate sanitation, garbage and waste collection), and reasonable quality of life issues (e.g., safe shelter, delivery of electricity, natural gas, telecommunication, public transport, personal security) [7]. There are many and varied deviations from municipality master plans for expansion that can affect urban redevelopment. Master plans are continually updated according to perceived social and economic (development) needs. Any such changes in a master plan will require citizen adaptability.

4.5 Potential Changes in Core City Elements and Outskirts to Protect the Health and Improve Living Conditions for Existing Populations

4.5.1 Spacing

Spacing and scale of enterprises for existing city cores and immediate surrounding, plus existing populations and projected growth in peripheral urban groupings and industrial ones distanced downwind from a city outskirts is a good starting point in redevelopment planning [7]. We previously described an ideal disposition of a city make up in a master plan for growth. As given earlier, this would be: (1) safe locations for residential housing (high cost, lower cost); (2) proximate open space (parks and recreational areas); (3) infrastructure to service society (e.g., government [education, public health, police, firemen, garbage collection, building code norms/updates]); and (4) utilities (water, electricity, natural gas, telecommunications) and transportation (roads bridges, bus/subway system, railroad, airport, port). In addition, commercial shops, markets, and offices can be mixed in with or close to residential zones, but with manufacturing, light industry away from and downwind of populations. Heavy industry, and waste disposal sites (dumps) would be located some distance downwind of populations and away and down flow from waterways that serve citizens and their activities. However, what has been reality is that a preferred spacing of the elements that make up many of today's highly populated cities beyond a colonial center core has not often been followed, especially in Asia, Africa, and South America. This makes rehabilitation of a city to serve all of its citizens now and in the future a formidable task that requires careful detailed planning to displace the minimum number of citizens while maintaining city activity as rehabilitation/redevelopment projects to improve citizens' well-being move forward in phased stages with the assurance of long-term financial commitment.

4.5.2 *Examples of Hazards and Approaches to Solutions*

Prevention of hazards is a challenge that has yet to be met for most hazards except for specific diseases for which vaccines have been developed and applied (e.g., smallpox, polio, measles, yellow fever). However, impacts from natural and anthropogenic hazards that endanger lives and physical assets in communities, especially those with high and dense urban populations, can be greatly mitigated [8]. Mitigation is based on physical, chemical, biological, and social solutions. In cases where the capability of predicting the onslaught of a hazard is possible, populations can be alerted and evacuated to safety although property may be lost. Property can be replaced, but lives can not. Example of solutions that have lessen risks to people from hazards are discussed in the following paragraphs.

4.5.2.1 Air Pollution

Whichever hazard negatively impacts the public health status or living conditions in a city core will certainly also affect the peripheral populations but possibly to a greater degree because of the shantytowns they inhabit and their health status. The hazard can be air pollution (e.g., with <2.5 μ particulates, heavy metals, toxic chemical aerosols and gases) that seasonally may hang over a city as a dense, sickening or killer smog. Among other things, it may also be seasonal flooding and links to diseases carried by sewage overflow, or stress from often clogged transportation routes during work days.

The sources of components that cause air pollution over major cities are many. They include juxtaposed industries upwind from cities and that lack state of the art pollution control technologies or they have the technology but fail to maintain or use it in order to maximize economic gain at the expense of spewing toxins into the atmosphere. These industrial pollutants are complemented by vehicular emissions from internal combustion engines and in some cases by the combustion of pressed soft coal used for heating during winter months and for cooking year round (e.g., in areas of China). As pointed out earlier in the text, a cloud of smog hanging over a city does not distinguish between people in economically advantaged neighborhoods or less advantages people occupying the shantytowns on a city's outskirts. All citizens are at risk if they engage in outdoor activities during smog conditions. Wild fires started naturally by lightening, by spontaneous combustion, or by arson, can pollute the air temporally over widespread areas but here, the fire is often the main hazard to property and infrastructure to a greater degree than air pollutants to people. "Controlled" seasonal burning that contributes to localized and regional "bad air" conditions is discussed in a later paragraph.

In some cities, industrial operations (e.g., battery factories, paint and pigment factories) are located in residential areas and continually poison their surroundings with air pollutants such as heavy metals (e.g., lead [Pb]) that have a great negative impact on the health of pregnant women, babies, and young children. Countries with aging populations that do not protect today's children from pollutants that can hinder their cognitive development (e.g., from lead poisoning), will suffer from the loss of educated, intellectually alert workers in the future. China is an example of a country that has responded to protesters factual concerns to begin to deal seriously to alleviate air pollution dangers to citizenry, especially the young (the future work force), while maintaining economic growth that contributes to the national GDP, albeit at a lower rate. Part of the enforcement has been to increasingly deal with corrupt officials that accepted or extorted bribes to turn a blind eye that allowed factories to spew out air pollutants.

Solutions to these health dangers exist but may be limited by government policies and short-term economic considerations without consideration of long-term economic/social/political costs. We have already considered what could be done to greatly reduce or eliminate industrial operations in the outskirts and upwind of cities that contribute large masses of toxic chemicals and fine-size particulates to seasonal smog conditions and/or heavy metal aerosols: the industries can be shut down during an at risk season; the factories can be moved to industrial parks downwind of existing and projected urban centers; the industrial plants can be outfitted with the best available technology of chemical scrubbers and particulate precipitators and certainly can be both moved and outfitted with pollution control systems. This is treated in some detail in a following chapter.

The vehicular exhaust input to smog conditions can be reduced by limiting the number of internal combustion engine powered vehicles that enter an urban center daily during an at risk season. As previously noted, this is done in many cities by permitting entry of vehicles with even number license plates 1 day and those with odd numbered ones the next day. Obviously, a redesigned road system that eases the flow of traffic and reduces the time in stalled traffic will lessen the mass of exhaust fumes released to the urban air. This can be planned for during city rehabilitation/redevelopment. Hybrid vehicles reduce the exhaust emissions to a city air and an increased use of electric vehicles puts no toxins into an urban atmosphere. It is likely that a reasonably priced (requiring subsidies) modern transit system with subways, surface trains, and electricity propelled buses and trolleys would greatly reduce vehicular traffic and its toxic producing emissions from internal combustion engines. This should be included in city redevelopment plans as something to be created or upgraded.

"Managed" seasonal agricultural burning such as slash and burn to create palm oil and pulp/paper plantations in Indonesia supports the "brown cloud" that pollutes air in South East Asia, (deep into Cambodia, Laos, Thailand and into Singapore and Malaysia) as monsoon winds move the "brown cloud" killer smog from Indonesia over the region [9]. The "brown cloud" extends over other Asian regions with input from industrial activity as well as agricultural burning in China, India, and other nations. The "cloud" is comprised of soot, fly ash, mineral dust, organic particles, sulfates, nitrates, and gases such as methane, carbon monoxide, and VOCs. Research indicates that the "cloud" abets global warming and the inhalation of the toxic air contributes to premature deaths [10, 11]. Burning of agricultural wastes (straw and stubble) from rice crops upwind of Delhi carry pollutants to Delhi and contributes seasonally to the industrial/vehicular driven smog that envelops the city. The burning is easier than uprooting and carrying away the agricultural wastes. To do away with seasonal burning of crop wastes, farmers can be educated to mix wastes into the soil to replace nutrients or to compost wastes that can do the same to support growth of future crops. This nutrient replenishment method can reduce the need for as much chemical fertilizer that would otherwise be used and be economically advantageous. A voluntary end of open burning of agricultural wastes will relieve previously affected populations of the burden of air pollution. If a farming community continues the open burning, politically strong national governments with the will and as guardians of their citizens health, can legislate against open agricultural burning and thus help clear urban atmospheres. There are also plans to consider use of the agricultural wastes as stock for biofuel production [12]. Wood burning and use of soft-coal (brown coal) for heating homes during a fall/winter smog season adds to the urban air pollution. Here, as before, municipal or national governments doing urban redevelopment have a social obligation to plan to bring subsidized fuel oil, natural gas, or electricity to populations that do not have them, thus reducing air pollution from these sources.

It is clear that to reduce urban air pollution that can engulf city cores and serviced neighborhoods as well as outskirt shantytowns with their economically disadvantaged populations will require concerted efforts. However, doing so will reduce premature deaths abetted by air pollution world wide. Such projects should be supported by committed economic resources from governments that choose to invest in the present and future. This may be supported by other players such as the World Bank and regional development banks offering low interest, long-term loans, and by grants from developed industrial nations and agencies they support such as USAID and like offices in many economically advantaged nations.

4.5.2.2 Flooding

Seasonal flooding affects many highly populated cities. The urban areas impacted most are determined by location and topography. Floods do not distinguish between locations where economically well off populations live and those housing low income populations. When an urban center suffers damaging, injuring, killer flooding from short term deluges or long duration rains, the lack of or failure of flood defenses often means that poor edge-populations, especially those in low lying areas, will likely suffer more so.

The costs to mitigate flooding that can injure and kill citizens and damage or destroy physical assets (e.g., buildings including housing, municipal structures, manufacturing and industry, utilities and transportation infrastructure) will be high if flood control greater than that for the so-called 100 year flood is not built up. This might be perhaps to mitigate the effects of the reach of a 500 or 1000 year flood or a 10,000 year flood such as that opted for by the Netherlands. The impacts on populations without creating or reenforcing flood defenses can be much, much higher than readapting cities to move waters quickly through them to spillways that can absorb or hold waters for later release. The Organization of Economic Cooperation and Development (OECD) released a report on potential increases in economic losses from exposure to damage and destruction in coastal city populations as a result of global warming (rising sea levels) and climate change (more intense, more frequent, and longer lasting storm systems) and surges [13].

For example, the OECD report sets New York City assets now at risk at US$320 billion, but projects that by 2070 (~2 generations) the at risk assets may be US$2 trillion. Five years after the storm Sandy that flooded some subway stations as well as parts of New York City, the last station was reopened in at a cost of US$346 million to rehabilitate it. Guangzhou now has risk assets at US$84 billion now, but by 2070 is projected to have have more than US$3 trillion worth of at risk assets, a little less than Miami. Other cities with most at risk assets now but that could increase greatly in the future are Tokyo, Amsterdam, Rotterdam, and Nagoya. The OECD and Asian Development Bank estimates that by 2070 the at risk assets will be greater yet in Kolkata, Shanghai, Mumbai, Tianjin, Bangkok, Ningbo (China), and Ho Chi Minh City [13–15].

Economic losses from flooding can be devastating. In 2017, Hurricanes Harvey, Irma, and Maria did economic damage to the United States of $125 billion, $50 billion, and $90 billion, respectively, much of which was caused by flooding. The economic strength of the country, states, and cities and resiliency of people have allowed a good degree of recovery. Review of flood control defenses and other factors that cause human suffering and financial loss are being evaluated to determine how effects of similar strength future events can be mitigated by better zoning and flood defense planning.

In 2016, a flood in Bangkok cost the city >US$4.6 billion and Thailand US$45 billion, thus reducing its GDP. The degree of flooding is intensified by the fact that Bangkok is low lying close to sea level and is suffering subsidence because of over pumping of aquifer waters (see Sect. 3.3). The Asian Development Bank estimated that US$1 billion invested in Thailand flood defenses such as improved waterways and pumping capacity could reduce the reach of the areas flooded by 50%. Clearly, this has to be complemented by controls that limit discharge of aquifer waters to volume of recharge waters to reduce and finally stop subsidence in Bangkok and other coastal cities with similar flood risk from torrential rains, long lasting rains, and storm surges.

In addition to economic losses from flooding, there can be losses to important health research centers. In 2001, tropical storm Allison flooded Houston, Texas and caused flooding at several medical research laboratories causing loss of materials, specimens, and data. Subsequently, research facilities installed special flood infrastructure such as special doors and floodgates to hold back storm waters (at Rice University, MD Anderson Cancer center, and the University of Texas Health Science Center). Baylor College of Medicine that lost 60,000 breast cancer specimens during tropical storm Allison constructed a wall around the entire campus and had no losses during Hurricane Harvey in 2017 [16]. Protection against flooding of such facilities either for new cities or for redevelopment of existing cities must have a high priority.

4.6 Defenses That Can Mitigate Flood Impacts on Citizens and Property

The World Bank estimated that for every US$1 spent on disaster prevention, US$4–$7 would be saved in recovery costs [17]. These figures are in question because when generated they did not distinguish between hazard type, its scale, the geographic region where it impacted, the effects of global warming/climate change, and other parameters [18]. This not withstanding, it is true that the funds are not always available for disaster risk reduction in the amounts necessary for overall mitigation/prevention projects. Thus, prioritization on how best to use available economic (and human) resources is necessary based on population security and asset protection.

4.6.1 Flood Defense Planning Procedure

Limited financial resources to build flood defenses during redevelopment projects is the major problem for some of the cities/countries with the highest risks to economic development and to injury/death threatened populations (mainly in Asia). The selection of flood defense options begins with topographic analyses of areas where potential flood waters endangering high risk locations originate and their flow paths that can result in heavy flooding. Physically, this level of flooding would impact urban functioning by damaging or destroying buildings and infrastructure that support daily living and economic revitalization. Second is an analysis of the urban setting to establish where one or more flood defenses would protect most people and installations by redirecting a water flow so that the volume of potentially incoming flood waters can be moved quickly through at risk zones to spillways where municipal master plan show no prospects for development thus keeping people and property out of danger. Third is an economic analysis of which flood defenses, if emplaced, would be most critical and best serve societal interests and that can be implemented given the economic resources available. Identification is necessary of other sites that are less critical and that must wait for additional funding to be invested in their defenses. Benefits/costs ratios for a project can be calculated and give a monetary value from initial investment figures. If positive, the project is worth considering at least in the short term. The greater the value, the more interesting investment there will be. If negative it may be rejected as an economic investment possibility. However, whether positive or negative, decision makers should make cost benefit analyses as well that are adapted to include not only estimated long term monetary return but also the physical, social and political benefits that investment will bring, added to long-term economic gains.

Planning for the future, flood defense programs have to consider at least a 50 year projection that assesses how global warming and climate changes would require adaptations (at increased costs) of the "now" implementation of flood defenses. Additionally, when developing a flood control emplacement plan benefit/cost ratio and cost/benefit analysis, planning teams have to include triggered effects such as the fact that rains that cause flooding may also cause landslides. Thus, the topography proximate to flood defense locations under consideration and the geologic (rock) composition of slopes that may be part of the landscape have to be evaluated with respect to landslide potential from in-seeping rain that can destabilize sloping terrain because of added weight to the slope and decreased friction in the underlying earth materials. For example, a canal configured to carry water away to prevent flooding would lose that capacity if blocked by a landslide and cause local flooding upstream and perhaps dangerous flooding downstream when the blockage is breached.

Some industrialized nations have invested heavily to protect their at risk urban and rural populations and property from flooding caused by storm surges. Because 20% of the Netherlands is below mean sea level and 38% is below high water level, flooding from wind driven high tides and North Sea storms is a threat to populations and well established economic ventures. In the Netherlands, this has been countered to a good degree by a carefully designed system of dams, sluices, barriers (dikes and levees). Rotterdam, the busiest sea port in Europe, is protected at a cost of at US$4 billion in 1997 (US$6 billion in 2017 dollars) with massive sea gates that are electronically closed when high water conditions threaten the city population and its port from what would be a one in 10,000 years flood from North Sea storm surges. Amsterdam is similarly protected. This Dutch expertise in flood control in low lying population centers could be applied to flood prone Asian urban centers such as those discussed in another section of the book if economic resources are made available. The British built a flood barrier in the Thames river at a cost of £534 million (US$776 million) in 1984 that would protect London from flooding by a North Sea tidal bore. Today's cost would be US$2.25 billion. The predicted rise in sea level has prompted England to plan redoing the flood barrier to sustain rising sea level assisted higher tidal bores. The United States invested $14.5 billion to strengthen levees and seawalls to protect New Orleans from a Katrina level hurricane but only for protection against a 1% chance of such a storm in 100 years. This was likely an economically driven mistake in decision making given IPCC projections on sea level rise and increasing extreme weather events globally. Investment in protection against at least a 0.2% of such a storm in 500 years was warranted to keep New Orleans safer against future extreme weather hurricanes.

4.6.2 Economic Limitations and Flood Defenses Methods

Developing and less developed countries with major flood problems that disrupt their economies and paths to development, especially in major urban centers that drive development, have to prioritize how available national economic resources should be used. These funds may be complemented by financial aid as low cost/no cost loans from institutions like the World Bank and/or regional development banks serving Asia, Africa, and Latin America. For example, as discussed in the previous chapter, Jakarta and Bangkok have important sections of their cities subject to flooding from torrential rainfall, tropical storms and associated surges, and typhoons. This becomes more of a problem because of subsidence in these (and other) coastal or near coastal cities (see Sect. 3.3). The flood threat will increase in the future with sea level rise. One solution proposed was to move at risk portions of cities to higher ground but this would economically unrealistic. To move the threatened north sections of Jakarta to safer ground was estimated to cost US$220 billion (Sect. 3.3.2.5). To protect Bangkok from flooding caused by storm surges abetted by rising sea level, Thailand considered emplacing a 100 km sea wall across the Gulf of Thailand but the cost was estimated at US$14.3 billion, too great a burden for the country to bear according to government officials (Sect. 3.3.2.2). We can review the options for flood defenses that are effective in maximally at risk locations and that can be realistically put in place physically with acceptable levels of governmental economic support.

Redevelopment to reduce flooding in core cities, established urban surroundings, and peripheral edge neighborhoods will require displacing and relocating people and altering other city elements (e.g., existing canals, roadways, structures) to be able to move potential flood waters quickly out of urban environs into spillways or retention dams thus reducing the exposure of at risk populations and property to damaging, destructive floods. This is readily planned but can be difficult to accomplish because of limitations on economic resources. Several flood defenses have proven effective at locations previously battered by large scale flooding. These include improving existing waterways by cleaning them out and by deepening and/or widening and/or straightening their channels and by installing walls on the waterways banks. This will mitigate flood effects by increasing the capacity to move greater volumes of flood waters more rapidly out of urban areas for discharge down flow of populations, critical infrastructure, commerce, and industrial parks into areas where they will cause no harm. In New Orleans, USA, high volume pumps in the city are used to move potential flood waters into the Gulf of Mexico. These have to be regularly maintained and tested to avoid failure of one or more that would enhance flooding as happened during the hurricane Katrina. Levees and dikes are used world wide to keep rivers in their channels or as barriers in coastal areas to limit the effects of storm surges that can be abetted when seas are at high tides.

We emphasize again that these modified channels and added structures have to be regularly inspected, maintained, and altered as needed. The Netherlands spends US$1.4 billion annually for their maintenance. The levee sea wall that protects Shanghai from the effects of typhoons and storm surges is raised after the existing sea wall is topped by a storm surge to a height that will not be topped by a like future storm surge. As described in Sect. 3.3, the subsidence problem in many cities that exacerbates the effects of flooding is caused mainly by unregulated extraction of water from aquifers. This has to be stopped to achieve a balance between discharge and recharge. This can be checked by monitoring the aquifer water table and by legislation that requires that water wells be registered or operators of uncovered illegal wells be discovered and be fined heavily or face incarceration. As previously noted, redevelopment that includes repaving infrastructure surfaces should consider using pervious cement that allows water to seep through them recharging aquifers and at the same time reducing surface flow and the threat of damaging flooding. The mitigation of flooding in coastal population centers from extreme weather is exacerbated by rising sea level and the extended reach inland of storm surges. The storms that contribute to these factors are likely partially, if not mainly, the result of global warming/climate change. The warming of surface ocean waters puts more moisture into storm clouds. This likely results in more frequent storm events with increased precipitation and of longer duration. This portends badly for growing populations in coastal or near coastal cities but can be damaging to inland populations as well.

4.6.3 *Seawalls for Citizen Safety and Asset Protection for Densely Populated Coastal Cities and Mega-Cities*

Many highly populated coastal cities (e.g., New York City, Boston, Bangkok, Jakarta) are evaluating projects to build near shore steel reenforced concrete sea walls to protect them and their inhabitants from rising sea level abetting storm surges and flooding. Without defenses inhabitants could be exposed to grave risk and physical assets destroyed. Careful planning for a seawall, its construction, and subsequent maintenance is a very costly proposition and may be considered by some as a "short" term investment given that sea walls histories suggest that they begin failing after 50 years. Failure is caused by "long" term corrosion of the steel supports in concrete sea walls by ocean water with its sodium chloride and magnesium sulfate chemicals seeping into a wall causing a sea water-concrete chemical reaction that degrades concrete and contacts and corrodes the supporting steel [19]. Factors that affect the life span of a sea wall in addition to the chemical reaction include the repetitive contact with wave and current activity, scouring at the wall base by back swash erosion, changing tidal water levels, sea bottom slope to seawall, seasonal temperature changes, and exposure to sun and extreme weather. Sea walls can disrupt shore processes (e.g., sediment transport) and habitats that hurt fish reproduction and can reduce biodiversity that supports the ocean/sea food web. Thus, a planning team evaluating potential locations for a sea wall structure to protect major city populations needs to go through an exhaustive assessment of a location for the wall that minimizes corrosion/disintegrating processes while preserving environmental integrity. A critical feature of a well constructed sea wall is that it can be heightened to account for rising sea level, and as experience dictates, future increases in storm surge heights. A seawall can afford some protection from tsunamis if its height is based on documented historical tsunami events.

With respect to a sea wall effective life span, a suggestion was made to use stainless steel to reinforce the concrete wall. However, this would drive up the cost by 6–9 times the US$300–$500 cost per linear meter and was not thought to be feasible. With respect to how to increase the resistance of concrete in the marine environment to degradation, a focus is on sea walls built by the Romans that have endured sea water contact for 2000 years. How? In a 2017 report, Jackson and her research team believe that this is due to the mixture they used (volcanic ash and quicklime [from limestone]) to make the concrete used for sea wall construction (without steel reinforcement rods). The volcanic ash for the Roman sea walls came from one quarry in Italy that contained the mineral phillipsite. Contact of phillipsite with seawater resulted in a rare chemical reaction whereby aluminous tobermorite, an authigenic mineral, grew out of phillipsite. The tiny crystal blades of tobermorite acted as tiny micro-armor that filled in concrete pores and prevented infiltration of concrete by seawater and likely preventing fractures thus adding to the long-term strength of the Roman seawalls [19]. If this reactive durable concrete can be reproduced, its use can save billions of dollars in seawall construction costs worldwide by not needing steel reinforcement and by reducing maintenance costs.

4.6.4 Population Growth: A Signal to Build Flood Defenses Now

Cities/mega-cities at locations susceptible to high risk flooding will continue their natural population growth added to by in-migration and immigration as allowed by national laws. An example of a city with people and a national economy with high exposure to flooding is Ho Chi Minh City that now accounts for more than 20% of the Vietnam GDP. The city has a 2016 population of 8 million that is projected to reach 13.9 million in 2025 and 20 million in 2050, a generation and a half in the future [20]. There is no way that a population increase of this magnitude can be protected from damaging, destructive flooding that has impacted, continues to impact, and will impact Ho Chi Minh City in the future. Only a prohibition of in-migration by the Vietnamese government and attention to flood defense options can prevent exposure of additional millions of citizens to floods and future tragedy. Examples of other cities with significant population growth by 2050 and with high value assets at risk from major flooding include Kolkata that is projected to add from 1.9 to 14 million people to the 2016 population of 14.8 million, Guangzhou that is projected to add from 2.7 to 10.3 people to the 2016 population of 18.8 million, Miami that is projected to add 2 to 4.8 million people to the 2016 population of 5.5 million, and New York/Newark is projected to add 1.5–2.9 million to the 2016 population of 8.8 million [14]. It is possible that the low end estimates of population growth for some of these and other major cities can be absorbed and protected from flooding if the planning for and building of defenses is initiated with some urgency given government will and access to economic resources. The cost of building coastal defenses for 136 cities is estimated at US$50+ billion annually through 2070 or almost US$3 trillion total. The OECD estimates that by 2070, without redeveloping for flood defenses against heavy deluges, tropical storms, typhoons, and sea level rise, US$35 trillion of world's assets in coastal cities will be at risk [13–15]. The clear message is invest now or lose heavily in the near and the two plus generational future.

4.6.5 Health Concerns from Heavy Flooding of Urban Centers: Storm Water Management of Sewage Overflow

Major flooding can have a public health impact especially in densely populated cities. This would be as diseases that develop in cities when sewage carrying systems are overwhelmed by flood waters and discharge pathogens from human and animal wastes into waterways used by citizens. This can take place where sewer systems carry both storm water and sanitary sewage through a single pipe, a combined sewer system that is a common condition throughout the world. Here the solution would be found in storm water management that designs methods that deal with overflow from sewer networks by directing sewage with human and animal wastes away from waterways used by people to where it will do no harm. This might be by adding infrastructure such as pipes and canals that channel the overflow to rivers that dilute the polluted overflow and move it to oceans, or that channel it to oceans directly. It might be by leading the human/animal waste polluted overflow to a retention area (e.g., pond, rock dam) for subsequent treatment before release to an environment or directly to a treatment facility. The overflow might be to infiltration installations into porous, permeable subsurface rocks. In some cases, sewer system pipes might be enlarged to as to be able to carry what would otherwise be overflow for safe discharge or for retention and subsequent treatment. It would also diminish the degree of flooding but would be a costly solution [21].

Another tact to solve the problem of a combined sewer system overflow (storm water, sewage, industrial effluents) is sewer separation with one pipe for storm water and a separate pipe for sanitary (+industrial wastes) discharge. This should be used in new purpose built cities. Storm water overflow can move directly into waterways whereas the organic waste polluted water overflow can be directed to a treatment facility or stored in retention areas for later treatment before release onto or into terrestrial or aqueous ecosystems or recycled into the cleansed water network. To replace a combined sewer system with separate sewer pipes for storm water and for pathogen-bearing sewage and industrial pollutants would be very costly to incorporate into full rehabilitation/redevelopment of most established cities that suffer the problem. The exception would be if new neighborhoods were purpose built to replace shantytowns/slums on city outskirts and these were to have their own treatment facilities. If this were the case, separate water carrying systems each with their own overflow paths should then be used.

These are the options that can be applied to secure that overflow from a combined sewer system or separated one be treated to a water quality that is fit for human consumption and use and also not dangerous to ecosystems into which the overflow many ultimately discharge. This means that there must be sustainable urban storm water management as was recently discussed in a journal paper [22]. Basic to proposing management protocols is a knowledge and understanding of laws that may affect planning and that can involve political decisions by local and regional governments. Management teams must have a realistic appreciation of economic and technical restraints as may exist. The protocols for planning a sewer system for a new purpose built city or for the rehabilitation/redevelopment of an existing system in a high population city or mega-city with high citizen density are much the same. These include some that are followed when planning flood defenses such as the geologic and topographic characteristics of an area that could affect urban storm water/waste water flow/overflow Decision makers must factor into their planning a knowledge of the climate, seasonal variations it might bring, and certainly an understanding of how climate might change with continued global warming (temporal and spatial realities) and affect the volume of water delivered to a sewer system. For existing cities, planners must be aware of storm water control (flood defenses) that are in place and determine whether they may be modified to retain or move more water to reduce or eliminate overflow. They must also assess treatment facilities and whether they can be upgraded to clean greater volumes of water. Before plans for new sewer systems or redevelopment of existing systems are proposed for final approval, planning teams have to judge what will be or could be (vis a vis global warming effects) the consequences of their decisions [22].

An additional potential public health problem related to sewage overflow would be the movement of persons in a densely populated city who may be carriers of an infectious disease contracted from contact with bacterially contaminated flood waters. A carrier can spread disease among the general population (e.g., cholera, intestinal ailments). Flood defenses that prevent overflow of sewage into waterways and a well staffed and provisioned public health system can help prevent an epidemic if a flood driven infectious disease problem does develop.

References

1. South China Morning Post. (2017, June 26). In China, Xi Jinping's new mega city is expanding is expanding underground. www.scmp.com/news/china/policies-politics/article/2099987/parts-xi-jinping-dream-city-be-built-underground
2. Cendrowski, S. (2017, April 28). China's newly imagined "green" city gets its first $19 billion in funding. On line. fortune.com/2017/04/28/china-xiongan-new-area/
3. Astana, Kazakhstan. Online. www.revolvy.com/main/index.php?s=Astana%20Kazakhstan&item_type=topic
4. Bogunovich, D. (2016). Urban development and planning inn the age of Megacities: An overview of current trends and current practices. Synthesis Report. Int'l Soc. of City and Regional Planners (ISOCARP), 33p. Online. https://isocarp.org/app/uploads/2016/06/MUF_FINAL_SYNTHESIS_REPORT_Aug2016-1.pdf
5. United Nations. (2015). *The millennium development goals 2015 report*. New York, 72p. Online. www.un.org/milleniumgoals/2015_MDG_Report/pdf/MDG%202015520REV%20(JULY%201).PDF
6. Holtz, M. (2017, December 7). Beijing goves migrant workers their marching orders. Christian Science Monitor, Online. www.csmonitor/World/Asia-Pacific/2017/1207/Bejing-gives-migrant-workers-their-marching-orders
7. Vorontsova, A. V., Vorontsova, V. L., & Salimgareev, D. V. (2016). The development of urban areas and spaces with the mixed functional use. *Procedia Engineering, 150*, 1996–2000.
8. Siegel, F. R. (2016). *Mitigation of dangers from natural and anthropogenic hazards: Prediction, prevention, and preparedness* (Vol. 127). Cham: Springer.

9. The Guardian (US edition). (2014, March 22). Indonesia's forest fires feed "brown cloud" of pollution choking Asian cities.
10. Scripps Institute of Oceanography. (2001). More knowledge of interaction between Asian brown haze, global warming ozone urgently needed. Online. http://www.scripps.ucsd.edu/news/2746
11. Sharma, S., Nenez, L., & Ramanathan, V. (2016). Atmospheric brown clouds. In: *Oxford research Encyclopedia of environmental science*. USA: Oxford University Press, 27p. Online. https://doi.org/10.1093/acrefore/9780199389414.013.47.
12. Damodaran, H. (2017, November 10). Delhi air pollution: A (crop) burning issue, and the way out. *The Indian Express*.
13. Nicholls, R., Hanson, S., Herweijer, C., Patmore, M., Hallegatte, S., Corfee-Morlot, J., Chateau, J., & Muir-Wood, R. (2008). *Ranking port cities with high exposure and vulnerability to climate extremes: Exposure estimates*. OECD, Environment working paper No. 1. Paris: OECD Publishing, 63p.
14. Stutz, B. (2012). Too big to flood? Megacities face future of major storm risk. Yale Environment 360. Unpaginated.
15. Hallegatte, S., Green, C., Nicholls, J., & Corfee-morlot, J. (2013). Future flood losses in major coastal cities. *Nature Climate Change, 3*, 802–806. https://doi.org/10.1038/nclimate1079.
16. Marris, E. (2017, August 31). How labs are coping with Hurricane Harvey's devastating floods. *Nature News*.
17. United Nations General Assembly. (2011). /11048 DP23. Online. http://www.un.org/News/Press/docs/2011/ga11048.doc.htm
18. Shreve, C. M., & Kelman, I. (2014). Does mitigation save? Reviewing cost-benefit analyses of disaster risk reduction. *International Journal of Disaster Risk Reduction: Part A, 10*, 213–235.
19. Jackson, M., Mulcahy, S. R., Chen, H., Li, Y., Li, Q., Cappelletti, P., & Wenk, H.-R. (2017). Phillipsite and Al-toberorite mineral cements produced through low-temperature water-rock reactions in Roman marine concrete. *American Mineralogist, Supplements, 102*, 1435–1450.
20. Ho Chi Minh City Economy. (2014). Online. www.alotrip.com/about-vietnam-economy/ho-chi-minh-city-economy
21. US Environmental ProtectionAgency, Office of Water. (1999). *Combined sewer overflow fact sheet: Sewer separation*. Washington, DC, 7p. Online. www3.epa.gov/npdes/pubs/sepa/pdf
22. Barbosa, A. E., Fernandes, J. N., & David, L. M. (2012). Key issues for sustainable urban stormwater management. *Water Research, 46*, 6787–6798.

Chapter 5
Planning for Existing Polluting Industries and Parks Zoned for Industrial Development

Manufacturing and industrial parks, and businesses that sell them materials and services, provide employment and community benefits for citizens within urban centers or from nearby populations. Additionally, workers, manufacturers/industries, and businesses pay taxes to governments that are best used to serve citizens' needs (e.g., maintain city infrastructure, social services) as well as municipal and state/national obligations (e.g., provide security for citizenry).

5.1 Characteristics of Light and Heavy Industries

Earlier in this book we reviewed the most desirable spacing for the different components of a newly built city. Within this were the locations for light industries that could be near residential zones and heavy industries (and waste dumps) a good distance away from and down wind of and down waterway flow from a core city and its planned neighborhoods for growing populations. Light and heavy industries have different characteristics in their physical footprint, skill needed by employees, toxins they may produce, water and energy requirements (e.g., electricity) and proximity to good transportation for workers plus transport for receiving raw materials and delivering commodities produced. The International Standard Industrial Classification of All Economic Activities gives a detailed list with all sub-categories to the major ones [1].

F. R. Siegel, *Cities and Mega-Cities*, SpringerBriefs in Geography,
https://doi.org/10.1007/978-3-319-93166-1_5

5.1.1 Light Industry

Light industries are small operations compared with heavy industries and generally require less capital investment and are more labor intensive. They locate near good transportation to receive small amounts of light raw materials to produce consumer goods. They are mainly near residential areas and major markets to easily move products to consumers. Light industries labor includes all manufacturing steps from processing raw material to assembly of products taking place most often in a single factory, generally with a small footprint. Consumer products from light industries include foods and beverages, home electronic appliances (toasters, computers), items for personal and home care (cosmetics, cleaning goods), apparel (clothes and footwear), furniture, textiles, paper, and plastics [2]. The fabrication processes of light industries can produce toxins but these can be readily captured with best available pollution control equipment tailored to each industry. Captured pollutants may be recycled or disposed so as not to put people or ecosystems at risk of pollution.

5.1.2 Heavy Industry

Heavy industries have high initial investments and maintenance costs. They are large operations occupying multiple buildings with infrastructure elements spread over many acres (hectares). They have skilled employees that use heavy equipment (e.g., large machine tools) in adjacent structures that change large masses of raw materials via complex processes into large and heavy products. Heavy industries often sell goods to industries as a supplier rather than to the actual consumer. Examples of heavy industries are those that generate raw materials (mining), make heavy machinery, create energy (e.g., electrical, heat), fabricate steel, build subways and railroads (cars and locomotives), and make arms for national defense (e.g., tanks, ships, aircraft, missiles). They also produce chemicals and petrochemicals and plastics, refine oil into gasoline, build automobiles, farm and construction vehicles, aircraft, and ships for civilian use, and construct buildings including skyscrapers, dams, and industrial complexes. In recent decades, the aerospace industry has been building rockets for space exploration and to launch satellites with mixed purposes into fixed or changing orbits [3]. These heavy industries and others not cited produce huge masses of toxins that would pollute the environment if not captured on site and recycled or treated and disposed of securely. Uncontrolled escape of emissions and effluents and unsecured disposal of solid and liquid wastes in dumps will harm human and ecosystem health.

5.2 Dealing with Existing Industries

Industries in many countries contribute to health risks for the urban populace and threaten the sustainability of ecosystems that support human activities. The emission, effluent, and solid waste pollutants differ with each industrial operation. In general, however, industrial generated pollutants that compromise the global health burden in different combinations and amounts are similar (Table 5.1) [4]. These are fugitive dust with some bearing toxic heavy metals (e.g., lead, mercury, arsenic chromium, cadmium, nickel, cobalt, copper), volatiles/vapors with some carrying organic compounds, sulfur dioxide, and heavy metals. There are effluents, some with solvents, acids (e.g., sulfuric acid) and other chemical compounds (e.g., nitrate, cyanide), pesticides and herbicides, heavy metals, and pathogens. Solid wastes may include pollutants released into the environment by leaching from dump sites, while incineration of solid wastes can generate dust, heavy metals, and toxic gases/vapors.

Table 5.1 Examples of polluting industries, important contributors to the global burden of disease, and their key toxic emissions and effluents [4]

Industry	Toxic components in emissions, effluents, and pathways to humans
Used lead acid batteries recycling	Pb, As, Cd as emissions and fugitive dusts via inhalation
Mining and ore processing	Pb, Cr, Hg, Cd, As, cyanide, radionuclides groundwater, surface water, soils – food, drinking water
Lead smelting	Pb, Hg, Cd, Cu, As, Sb, SO_2, volatiles as fumes, dust particles, waste water, solid wastes to air, soil, water
Tanneries	Cr, Pb, acids, alkalis, biocides, animal flesh with pathogens from wastewater into soil and groundwater
Artisian and small gold mining	Hg, Pb, Cr as dust vapors, inhaled and ingested in Hg contaminated water, fish
Industrial/city dumpsites	All very different, Pb, Cr, d, As, other volatile metals, pesticides, hazardous organic compounds, pathogens As dust, leached into groundwater, and as VOCs
Industrial parks	Pb, Cr, inhalation of dust, ingestion of contaminated water, food (irrigation), LMICs
Chemical manufacturing	A myriad of operations that include inorganic and organic chemicals, paints, cleaning pharmaceuticals, cleaning products, inks, chips, explosives. Generate many byproducts in waste process water: solvents, dyes, pesticides, As, Cd, Hg, Cr, Pb, cyanide, via emissions (particulates, dust), accidental spills, bad waste disposal via inhalation, contaminated water, food
Product manufacturing	Pb, Cr, Hg, Cd, As, SO_2, cyanide, VOCs, as from heating during processing, as emissions, from incineration, to water, soil, solid waste disposal
Dye industry	Sulfuric acid, Cr, Co, Pb, Hg, Cd, S, As, Ni, Co, nitrates, chlorine compounds, ingestion via contaminated water and food crops irrigated with contaminated water, complexly varied according to dying processes used

Table 5.2 Worst polluting industries as measured by DALYS [4]

Industry	DALYS
Used lead acid batteries	2,000,000–4,800,000
Mining and ore processing	450,000–2,600,000
Lead smelting	1,000,000–2,500,000
Tanneries	1,200,000–2,000,000
Artisanal small gold mining	600,000–1,600,000
Industrial/municipal dump sites	370,000–1,200,000
Industrial parks/estates	370,000–1,200,000
Chemical manufacturing	300,000–750,000
Product manufacturing	400,000–700,000
Dye industry	220,000–430,000

In many countries, where industrial/manufacturing output is a basis for economic development, pollution controls are strict. However, in others, especially in South East and South Asia, Africa, and South and Central America, it may be lax and contribute to the global health burden that can be measured as DALYS (disability adjusted life years) or the number of years lost due to ill health, disability, or early death. This is the result of inhalation or ingestion, over time, of toxic air, polluted water, and contaminated food, at the workplace, on the street, and in the home. It has a direct affect on a country's GDP because of a loss of productivity when people are no longer able to work and contribute to economic development. DALYS attributed to 10 important industries are listed in Table 5.2 [4]. A great number of the industries listed in Table 5.1 use coal as their energy source. Without effective and efficient pollution control technology as a final step in their operations, they emit fine size particulates and heavy metals added to their CO_2 contribution to global warming/climate change.

5.2.1 Options to Thwart Health Threats from Urban Industrial Pollution

For existing manufacturing/industrial sources of air, water, and solid waste pollution, in the midst of a large population center or upwind and/or up flow of it, there are three options to ease or eliminate pollution emissions, effluent discharges, or solid waste disposal problems that were described in Chap. 3. As given in Sect. 4.5.2.1 and further discussed in the following paragraph, one will result in loss of jobs, one that may not be economically feasible, and one that is financially more acceptable but may not be socially acceptable without legislated and enforced controls on emissions, effluents, and disposal of solid wastes.

The first option is to shut down a facility if the owners do not want to consider one of the two options given below. If this is the decision, then environmental laws should be on the books that requires facility owners to remove and securely dispose of a facility's inventory of potential or active pollutants (gases, liquids, solids) including those that may be stored in underground tanks. The second option is to move the facility to a location downwind and down waterway flow and with the requirement that it must be fitted with pollution capture and control equipment. In this way it will not expose municipal master planned future neighborhoods or a new population center to pollution in future years. The expense in relocating an industrial facility is substantial because the infrastructure has to be in place: utilities (gas. electricity, water, sanitation), roadways, transportation to move workers from their urban location to and from their places of employment, and perhaps a railroad spur or port to receive raw materials and to export product. Whether a municipality opts to invest in necessary infrastructure and/or provide limited tax concessions to encourage an industry to move a facility to new location depends on the long-term benefits/costs ratio as well as a benefit cost analysis that includes socio-political factors that might favor or negate this option (e.g., keep employment, tax income, domestic spending). The third option previously proposed is to install the best available pollutant capture and control equipment in an existing facility that is a source of pollution but not located in the midst of a residential or commercial zone. As markedly improved emission control chemical scrubbers and electrostatic particle capture equipment are manufactured, an industry should be required to add to or replace existing equipment. The latter two options would require the treatment of effluents to remove pollutants and the secure disposal of generated solid wastes perhaps by their recycling within or by shipping them to an industry that will use them.

It is imperative that industries working with or that plan to work with hazardous materials and that are located at sites known to be at risk of being struck by a natural or human caused hazard protect the public and their employees from the toxins a hazard may release. Such releases have occurred in the past from an earthquake, a tsunami, flooding, turning a wrong valve, not giving immediate warning for evacuation. For example, not sounding immediate alerts calling for evacuation when the Chernobyl and Fukushima nuclear facility events occurred caused many deaths and subjected many proximate and distant populations to an increased risk of radiation poisoning from the release of radioactive matter over extensive areas. The results of the tsunami driven disaster at Fukushima will not be relieved for many years in the future both on land and in the Pacific Ocean where radioactive water continues to leak from the damaged reactor to contaminate ocean ecosystems. This disaster could not have been totally avoided but as previously noted its effects could have been significantly damped by raising the "protective" sea wall by 3 meters or more as would have been indicated by a thorough geological study that was lacking but that was later revealed in a 1981 journal report [5]. A heightened seawall would likely have reduced the inland run up and force of the tsunami wave preventing some loss of life and property. It may have prevented the damage and invasion by sea water to backup generators by the tsunami driven sea water that entered surface exposed structures and shorted their electrical system so that there was no power to keep coolant moving through the Fukushima reactor cores. I have not read of any modeling done that supports or denies the effects of a raised sea wall as regards the Fukushima event.

It is interesting to note that in the state of New York, USA, the Indian Point nuclear power facility, 38 mi (54 km) north of Manhattan, New York City will be closed down by 2021. The plant has 3 reactors and was built to withstand an earthquake with a Richter magnitude of 6.1.The reasons given for its decommissioning are because of its age (40 years in service), 40 accidents (e.g., transformer explosions) and releases of radioactivity (e.g., leakage of tritium contaminated water into the groundwater at the facility in 2016), but also because a seismic zone (potential earthquake activity) discovered in 2008 is located one mile north of the facility [6]. The Fukushima event in 2011 likely influenced the decision to decommission the facility.

Planning against natural hazards that could affect industries may be done first by avoiding sites that will put an them at foreseeable risk. Second, defense against some hazards can be implemented with some confidence by infrastructure changes, by retrofitting structures and infrastructure, by securing hazardous materials against release by an event, and by monitoring them to detect a toxin release so that a planned for response, perhaps citizen evacuation, can be initiated.

5.3 Planning for Industrial Development Proximate to a New City

Siting industry locations or industrial parks for a city to be built from scratch by governments with strong industrial ambitions can be a relatively easy problem to solve. Experienced city designer-architects will first determine the renewable water sources and what volume will be available for industrial use. This means that the number of persons that are expected to move to the new city over time has to be projected with some certainty to calculate the per capita amount of water needed to service their demand exclusive of industry requirements. This will allow an estimate of the water available to industries that would be attracted to a new population center. One key here that was mentioned in previous chapters is renewable water that means that aquifer waters cannot be extracted at volumes greater than the rate at which they are recharged. As described in Chap. 3, this obviates a potential subsidence problem and susceptibility to flooding that is afflicting major cities. Another key is if industries recycle water as in Japan, for example, where there could be a 90% recycling, thereby reducing the draw on renewable water supplies. A third consideration that has to factor into long-term adaption planning for city water security is a possible change in water supplies because of global warming and regional climate change.

Once the water supply capabilities are established, planners for a new city have to establish possible locations for industrial parks, the infrastructure needed to serve industries including energy sources and transport systems. This is done within a master plan that observes a spacing for all components of a new city that insures safety for urban dwellers from possible industrial pollution. As described in

Sect. 4.1, this includes exclusive and mixed residential areas, commercial areas, government structures in a center city and proximate planned neighborhoods with light industry on the outskirts, and with heavy industry and waste disposal sites well distanced from populations. A basis for siting city components has to consider the history of natural hazards (their frequency, force, areal reach, and duration) that has been recorded or as revealed by geologic and geographic analysis of the areas being considered for the building of a new city. If a reasonable location for an industrial park does have a threat from one or more natural disasters (e.g., flooding, landslides, earthquake), planning teams have to determine what could be done structurally or otherwise to secure an industrial installation from a recurring hazard. Damaging earthquakes are rare, floods are common as are landslides. There are building codes to follow, flood control methods that can be applied, and possibilities of reducing landslide hazards but protective methods are costly so that alternate locations may have to be considered.

5.3.1 Acceptable Industries and Their Plans to Prevent Urban Pollution

When a site has been approved as acceptable for industrial projects, the question is what industries are acceptable to the planned city and under what conditions that preserve the health of citizens and ecosystems. Each industry has its own raw material needs: gases, liquids, and solids. Where are raw materials for an industry stored…above ground or in buried storage tanks? How are these monitored to insure against a loss/leakage and intrusion into an environment. If there is a loss/leakage, what warning and alert system is in place and containment technology ready so as to protect workers and ecosystems (e.g., surface water or aquifers) that might be threatened by escaped raw materials? Specific industries generate their own emissions, effluents, and solid wastes and modifications thereof depending on differences in processing. Will a proposed industry have state of the art equipment that captures at least 90% of potential atmospheric and water borne pollutants they generate and is this enough given a volume of production? What are the plans for recycling captured pollutants, or treating and cleansing them to legally permitted levels before release into an environment, or preparing them for recycling or secure disposal? Solid wastes as may be generated as part of an industrial process can be recycled if this is economically beneficial to the industry. If not, is there a plan and place to dispose of them in secure landfills designed to capture leachate from rain water or melting snow seepage by directing it into a impermeable lined pit (or tank) for subsequent recovery and treatment? This will prevent interaction with people and a natural resource-rich environment, especially groundwater with a flow path towards an urban water source. Accidents happen at industrial sites because of equipment failure, human error, or sabotage/terrorism. Thus, as previously cautioned, monitoring systems should be in place to give warning and alert the public in the event of a unexpected release of toxins into the air or waterways that could reach an urban population a good distance down wind and/or down flow of pollutant release. If a licensing authority and industry executives agree to these regulations, proposed industrial park (s) locations can be put in a city development building plan.

5.4 Uncovering Potential Legacy Threats to Populations

When planning rehabilitation/redevelopment of an existing city that will not only improve the living conditions and quality of life for all residents especially those inhabiting shantytowns, the legacy effects or inheritance of conditions that affect terrain usefulness for development have to be assessed. This includes sources and effects of pollution from closed down and abandoned light/heavy industries as well as projects that have disrupted ecosystems and the biodiversity they contained. Evaluation of areas where citizens displaced from shantytowns may be relocated to new neighborhoods with proper housing serviced by sanitation and utilities has to take place as well.

5.4.1 Pollution

Legacy pollution includes what may have been left behind that would negate the reuse/new use of land unless reclamation makes it safe (clean) for human habitation and development ventures [7, 8]. This might be soils contaminated by heavy metals and/or inorganic and organic chemicals from industrial emissions that precipitated onto them or effluents that flowed over them depositing some contaminants and carrying some into waterways. It might be groundwater tainted by seepage of pollutants through soils or leakage from cracked sewer lines into aquifers. A knowledge of industries that were active in the past can provide guidance to the pollutants they generated. Municipalities have to assess as well what wastes industries may have left behind and if so, where and in what condition were these wastes disposed of? Were they just thrown in dug ditches and covered or buried, or encapsulated in metal barrels and buried? Also, were raw materials for an industrial process stored in underground tanks/vaults that were not emptied with their contents securely disposed of before leaving the industrial site? Information of this type that is essential to planners can come from municipal records, newspaper articles, and former workers or their families. Without documentation that gives answers to these questions, surveys can be made to establish the suitability of a terrain for development, These may be by soil and groundwater chemical analyses to establish the usability or degree of contamination. They may be by ground penetrating radar, magnetometers and other methods to reveal any buried (hidden) possibly toxic wastes. Such field (on site) studies should be carried out to certify that the land under consideration for use is safe for habitation or other use. Evaluation of surface/subsurface analyses can determine whether or not reclamation is necessary. If reclamation would be necessary, potential development investment will depend on benefit/cost ratio and cost benefit analysis.

5.4.2 Legacy Threats Other than Pollution

Projects that have disrupted ecosystems with their natural resources that may be important, directly or indirectly, to service urban redevelopment/rehabilitation, but that are not pollution related also produce legacy threats. These include river diversion (for irrigation, drinking water, industry), dam emplacement (flood control, reservoir, hydroelectric power), and vegetation removal (forestry).

5.4.2.1 Dams

Dams are built as reservoirs to secure water supplies for times of drought, as a means of regulating river flow for flood control, and to generate hydroelectric power. They can be important to city rehabilitation/redevelopment as growing populations require electricity supplies, protection from flooding, and available emergency water supplies. In 2015, 3688 dams were being built worldwide to bring electricity to 1.4 billion people without this commodity and also to reduce coal-powered energy generation, and for some, reliance on nuclear power facilities. Dams can create lakes that serve recreational needs and attract tourism but otherwise can cause ecological damage upstream and downstream [9]. Most are being built in Southeast Asia, Africa, and South America, regions that have most of the world's mega-cities and high population cities that have steadily increasing population growth mainly from natural births and demographic change. Although built to serve humanity, dams emplacement can have negative impacts on people and ecosystems that urban planners considering dam emplacement have to consider.

For example, dams block fish migration (e.g., salmon, trout, eel, and sturgeon) from where they have been feeding and growing in the return to their spawning grounds. This can lead to diminishing populations of food fish unless modifications are made that allow the migration such as ramps for salmon in the United States Pacific Northwest. Dams trap sediments (sand, silt, clay) and this harms downstream spawning habitats that disrupts ecological balance. Sediment trapping starves fertile floodplains, seafood productive deltas, and inshore wetlands of their natural supply of nutrient bearing sediment, thus threatening a source of food security. The loss of sediment load (behind a dam) results in scouring a river bed downstream and undermining of its banks as flowing water establishes a new equilibrium between its kinetic energy and the sediment load it can transport. If an aquifer receives recharge by inflow from river water, a lowering of a stream bed will lower the groundwater table and reduce an aquifer capacity to serve a population. Upstream, fertile land will be lost to rising water that can affect upstream flora and fauna in response to slow moving water. This may support health threats to populations especially in tropical regions where still waters may be breeding environments for mosquitos (e.g., vectors for malaria, dengue disease) and snails (e.g., carriers of flukes that cause schistosomiasis) as well as providing a new habitat for crocodiles. This rise of water behind a dam will cause the water table to rise and could cause a tragedy if the walls of the valley are composed of sedimentary rocks that will weaken as water seeps in increasing the weight stress and lubricating sediments thus increasing the effect of the pull of gravity. A landslide into a filled dam site can displace water over the dam causing flooding in the valley and downstream. This happened in Italy in 1963 at the Viaont Dam in the Piave Valley and caused the deaths of more than 3000 people. A planning team focussed on redevelopment of a city that benefits from an existing dam and urban expansion to accommodate population growth has to consider possible inherited threats as described that might be future problems. The assessment is crucial as well for a potential dam emplacement site with respect to population presence. One would hope that designers of the more than 3600 dams now being built have learned the lessons from problems with dams built in the past.

5.4.2.2 Water Mass Diversion

Diversion of rivers is a common happening to bring water for irrigation, for drinking and cooking (after treatment), and for manufacturing/industrial facilities. A classical case in point is the Aral Sea. At the beginning of the 1960s the Aral Sea covered over 25,000 mi^2 (~66,000 km^2), contained 1000 km^3 (240 mi^3) of water, and was fed by two rivers with 65 km^3 (~15.5 mi^3) of water annually. The USSR diverted the river water to irrigate cotton fields to the point that only 1.5 km^3 (<0.4 mi^3) of river waters were delivered to the Aral Sea annually. It shrunk 80% into two parts. What was a thriving fishery and popular tourist site became a salt-sand desert. Winds carried salt to farmlands over a radius of 300 km (~186 mi) to render them unproductive. Only recently has the World Bank loaned money to Kazakstan and Uzbekistan to try to reclaim the sea but this is a long term project. Similarly, during 35 years, waters that fed Lake Chad (bordered by Chad, Niger, Nigeria, and Cameroon) were diverted for irrigation that when combined with low rainfall over the years shrunk Lake Chad by 95%. These stories are repeated worldwide whether for California's Mono Lake that was shrinking because of drought so that water export to Los Angeles had to be reduced or for Mexico's Lake Chapala where water from the lake is not diverted but rather drained for drinking water for increased population (> 5 million people) in Guadalajara and for increased irrigation. Since the 1970s, Lake Chapala shrunk more than 20% with its sustainability in question unless controls on discharge vs recharge are instituted [10]. The legacy threats here are loss of ecosystems and much of the natural resources (e.g., drinking water, fisheries, food crop agriculture) they provide to populations nearby and at a distance. Planning city redevelopment that may provide river/lake/sea resources to an urban population has to determine what natural resources could be lost that otherwise could help sustain population growth if these water sources were not tapped.

5.4.2.3 Vegetation Removal/Forestry

Removal of vegetation by forestry projects, for expansion of a city, or for increasing agricultural terrain, represents encroachment and can leave their legacy threat as well. In place vegetation can sustain slopes against rain induced landslides in sloped areas and can also slow down and reduce volumes of rain-fed flowing water as a natural flood control. Thus, depending on an environment being considered (e.g., geologic, topographic, climate) for urban creation or expansion, and the loss of habitat and biological diversity, the vegetation factor has to be considered. Lastly, vegetation removal may be necessary as a fuel removal process against an ofttimes seasonal threat of an advancing wild fire.

References

1. United Nations Department of Economic and Social Affairs. (2008). *International standard industrial classification of all economic activities*. Revision 4. New York: United Nations, 291p.
2. Wikipedia. Light industry. Online. https://en.wikipedia.org/wiki/Light_industry
3. Wikipedia. Heavy industry. Online. https://en.wikipedia.org/wiki/Heavy_industry
4. Pure Earth and Green Cross. (2016). *World's worst pollution problems: The toxins beneath our feet*. Zurich, 53p.
5. Minoura, K., Imamura, F., Sugawara, D., Kono, Y., & Iwashita, T. (2001). The 869 Kogan tsunami deposit and recurrence interval of large scale tsunamis on the Pacific coast of Northeast Japan. *Journal of Natural Disaster Science, 23*, 83–88.
6. Wikipedia. (2016). Indian Point Energy Center. Online. https://en.wikipedia.org/wiki/Indian_Point_Energy_Center
7. James, L. A. (2017). Legacy effects. In *Oxford bibliographies, environmental science* (pp. 1). Oxford University Press. Online. www.oxfordbibliographies.com/view/document/obo-9780199363445/obo-970199363445-0019.xml
8. Hanrahan, D. (2012). Targeting legacy pollution. In *Getting to green – a sourcebook of pollution management policy tools for growth and competitiveness* (pp. 57–69). Washington, DC: World Bank. Online. siteresources.worldbank.org/INTERNETENVIRONMENT/Resources/244351-1279901011064/GuidanceNoteonLegacyPollution.pdf
9. Zarfi, C., Lumsdon, A., Berlecamp, J., Tydecks, L., & Tochner, K. (2015). A global boom in hydropower dam construction. *Aquatic Sciences, 77*, 161–170.
10. Larsen, J. (2005). *Disappearing lakes, shrinking seas*. Earth Policy Institute. Online. www.earth-policy.org/plan_b_updates/2005/update47

Chapter 6
Curtailing Agriculture Projects' Practices That Can Harm Urban Food Security and Public Health

6.1 Introduction

Many old and "modern" cities have been built on terrain that was previously vegetated and part of biologically diverse ecosystems. This was often on productive farmland that was not protected from encroachment as municipalities expanded and opened pristine areas to development. The encroachment continues as rural citizens and their families are attracted to urban centers for the way of life they seemed to offer. This starts with employment possibilities, education for workers children, and family health services as manufacturing/industrial operations and large and small businesses that serviced them spurs economic development. Agricultural projects are located mainly away from population centers, be they major metropolises or smaller cities. They provide food security for urban citizens through "brought into markets" in-ground crops, bush crops, and tree crops they cultivate, and through animal husbandry (e.g., beef and dairy cattle, poultry, hogs, sheep). The last chapter described how manufacturing/industrial endeavors can present potential harm to the health of urban and rural citizens and to natural resource rich ecosystems that help sustain them, and how to find solutions to the problems. This chapter will discuss the problems that food production methods present, the threats they pose to urban populations that are generally distant from them, and possible solutions to such problems.

It should be noted that animal husbandry range stock will cause little environmental harm if herds and flocks are properly tended. However, commercial high production operations can create massive waste-related problems (e.g., manure, urine, pathogens [e.g., salmonella, E. Coli, listeria, coronavirus, foot and mouth disease]). Added to this are the wastes of blood, innards, bones, heads, feet, skin, feathers from the slaughter and dressing of ruminants and poultry. In addition, there are the problems from the overuse of antibiotics and growth hormones (anabolic steroids) delivered to food animals via feed or patches. The overuse of antibiotics for animal husbandry may stimulate the mutation of target bacteria to antimicrobial

F. R. Siegel, *Cities and Mega-Cities*, SpringerBriefs in Geography,
https://doi.org/10.1007/978-3-319-93166-1_6

strains that can resist antibiotics treatment of human bacterial infections and result in patient death. The risk to human health is a worry to the medical community [1]. World wide it was estimated that in 2014 there were 700,000 deaths attributed to microbial resistant infections [2, 3]. This problem can begin to be alleviated by using antibiotics only for sick animals and not an entire herd or flock and research that focuses on medicines to kill the existing antibiotic strains. This should be coupled with the elimination of the use of growth hormones in commercial operations where food animals or food-giving animals are raised. Residues of these antibiotics and growth hormones have been found in the meats from food animals and over time can build up in consumers (billions in cities) and be dangerous to their long term health. We will discuss what governments are doing to minimize the use of chemicals in animal husbandry operations.

6.2 Food Insecurity: Stemming Urbanization

Many in the global population, especially babies, young children, and the aged in Africa and the Middle East suffered with chronic malnutrition and starvation from famine in 2017. This is the result of drought, people being displaced from their homes by warring factions and decisions not to allow humanitarian food supplies to reach starving populations, thus exacerbating the famine problem. If there is agricultural land that adds to the food security and it is nearby expanding cities, this land should be preserved and neither encroached upon by growing populations nor be subject to seizure by eminent domain laws for economic development. **Here we have a moral question: Should cities limit population growth from in-migration and immigration to prevent encroachment?** The flow of people to cities can be arrested if urban attractions are not driving forces. This means improving the services available in rural areas by subsidizing farming, increasing employment opportunities, and upgrading and increasing educational offerings. There should also be improvement in rural health care facilities with professional and support staff, and well stocked with basic medications, and with ready access to diagnostic medical equipment. This solution is easily prescribed but difficult to bring to fruition mainly for economic reasons but also by the reluctance of some trained personnel to move to rural areas such as those needed to provide professional health services. Nonetheless, a start in bringing this prescription to reality may slow urban population influx from in-migration and lessen the possible encroachment on productive land.

6.3 Erosion Control and Water Supply/Needs

As indicated above, it is good practice to preserve farmland near an urban center not only from population encroachment but also from processes that can degrade good agricultural soil and affect the quality and yield of food crops grown and hence food security. The former action requires a municipal decree that will prevent squatting and establishment of shantytowns on protected land. The latter requires that consultants assess erosion potential of farm acreage and provide solutions to prevent erosion by wind (e.g., change plow and seed method to slice and seed; plant vegetation wind breaks at peripheries of croplands) or by flowing water after heavy rains (e.g., by diversion channels to safe discharge locations) if this could be a problem. When protected land is to be used for agriculture, experts have to determine what water requirements will be for crops that are planned and consider how this might change if crops are rotated (alternated) and thus affect urban water security. Will crops be irrigated or rain fed, or be sustained by a combination of both? If irrigation is necessary, will the irrigation technique to be used preserve water supplies (e.g., drip irrigation that needs initial investment and subsequent maintenance) or will it lose precious water to evaporation during arm irrigation? Will the irrigation water come from aquifers or surface waterways? If from aquifers, is the seasonal discharge balanced by recharge so as to sustain water supplies and prevent subsidence as may occur? If subsidence were to take place, there can be additional problems that affect a population that were discussed in a previous chapter (e.g., greater susceptibility to flooding, ruptured infrastructure). In planning for the future there is the question of how will projected climate changes affect availability of the agricultural water supply over time and perhaps require changes in crops that should be grown?

6.4 Agricultural Chemicals and GMO to Assist Growth and Protect Crops

Food security for citizens from farm crops can be abetted by the use of agricultural chemicals (fertilizers) that replenish the nutrient content of soils depleted by a previous crop. The application of herbicides protects crops from weeds that take up nutrients while pesticides repel pests (insects, animals) that infect or consume crops. The judicious use of an optimum amount of pesticides, herbicides, and fertilizers on in-soil, bush, and tree crops will assure full growth and maximum crop yield and limit their effects on any excess that may access ecosystems. That there is no overuse is essential. It is also economically advantageous to a farmer by saving on agricultural chemicals expenses. Overuse of these chemicals can result in runoff of an excess into nearby surface waters when it rains, or after irrigation, and can also result in seepage of the chemicals and biocides into aquifers thus contaminating groundwater that may flow towards cities and be used by urbanites. Contaminated groundwater that discharges into waterways can affect the life forms inhabiting them to the detriment of society by harming an important food source. For example, runoff of nutrient rich agricultural chemicals into waterways will contribute to algal blooms that can be responsible for great fish kills (red tide) in oceans, estuaries, lakes and rivers as phytoplankton comprising the bloom die and release toxins that attack fish gills. Red tide also causes closure of shellfish beds for harvest and may cause skin problems and respiratory distress in people in coastal zones from wind borne toxins. As important is the fact that decomposition of dead organisms (mainly algae) in marine and terrestrial water bodies can use all of a habitat's oxygen (eutrophication) thus causing fish kills and dead zones. Such actions reduce the yield of food fish for cities and for rural coastal populations.

6.4.1 Genetically Modified Foods

Genetic modifications can produce seeds that may protect farm crops from weeds and repel pests that harm crops or diminish yields thereby avoiding the application of chemicals to do the job. This eliminates chemical residues on vegetables and fruits, that if not thoroughly washed, can enter the human food chain and potentially be harmful to consumers. However, ingestion of GMO grown products could be a risk to human health as suggested by controlled experiments. In these, rats developed lesions, suffered organ damage, and died after consuming GMO grown potatoes. However, a critical review of the literature suggested that the published research left an incomplete evaluation regarding the safety (toxicity) of GM crops eaten by humans and fed to animals. This is because there was no unified experimental methods so that reports could not be properly reviewed, and perhaps more important, experiments could not be repeated, a key to validation of experimental results [4]. There have been failures in GM grown corn that contributes to mutated, resistant insects that attack GM growing crops and pesticide protected crops. For example, the rootworm that infected corn developed a resistance to GMO corn treated with Bacillus thuringiensis (BT) [5, 6]. In a sense, this can be likened to the overuse of antibiotics in cattle feed consumed in feedlots that resulted in the mutation of bacteria to strains resistant to antibiotic therapy. There is the fear that during pollination of GM crops, the pollen will be carried to non-GMO fields and affect the purity of their crops. In Capulalpam, Mexico, scientists documented the discovery of transgenic genes in the Mexican native corn. In 2013 Mexico banned the entry of GM corn. It is because of uncertainty of their effects on human and animal health and food security that GM crops are not allowed in the European Union, several nations in Africa, and elsewhere. In addition, some researchers believe that there is the potential for an allergic reaction in humans from ingestion of GM based corn products. Organically grown foodstuffs obviate the chemicals or GMs problems but do not produce the mass of crops needed to feed large and growing global and urban populations as was the case for the GMO "green revolution" that allowed many countries to improve their food security.

6.4.2 Microbes to Protect Crops and Increase Yields

Microbes associated with plants from biomes on all continents except Antarctica have been studied in great detail and some identified as to the characteristic they give to plants that assist agricultural projects. These have been cultured, batch produced, and used to coat seeds. The first product of this effort, Indigo Cotton™, cotton seeds coated with a proprietary mix of microbes (bacteria and fungi) were sown in West Texas in 2016. This resulted in an 11% improvement in yield by protecting the crop against drought stress [7]. Similarly treated seeds of soy, rice, wheat and corn coated with proprietary microbe mixtures have been produced with like results. This adds to the food security of city and mega-city populations dependent on foods from outside the urban locations, complementing advances in natural hybridization and GMO that impart seed properties that protect crop yield. The company is working on microbial products that can reduce the need for nitrogen fertilizer and that offer protection against pests.

6.5 Animal Husbandry Wastes: A Threat to Public Health and Critical Ecosystems Unless Used, Controlled, or Disposed of Securely

The demand for food is rising worldwide with increasing populations, and where incomes are also increasing, the demand for meat is rising as well. For example, the Chinese population increased by more than 40% from 1980 to 2015 and the income has increased significantly more. This resulted in a fivefold increase in the demand for pork. These changes put pressure on the animal husbandry industry to produce more product, ofttimes at a faster rate, using chemicals and vaccines. The problems that exist in this industrial production mode and solutions to them are discussed in the following paragraphs.

6.5.1 Animal Husbandry Problems that Affect People

Another major food security concern for growing urban (and global) populations are problems with animal husbandry during growth and after slaughter and distribution. It is most threatening to ecosystems close to where thousands to millions of animals are bred in close quarters (e.g., cattle feed lots and commercial chicken farms, respectively) and where they are taken for slaughter. The problems are how to capture and control the wastes and how to safely use them or dispose of them so as not to contaminate soils and waterways or aquifers. These are manure and urine from cattle (plus pigs, sheep, and goats) and fecal matter from chickens (plus ducks, turkeys). Added to this are slaughter house wastes (fluids [blood, wash water]) and solids (heads, innards, skins, bones, feathers), plus pathogens (e.g., salmonella, E. Coli, listeria, coronavirus, foot and mouth disease) from dressing the food animals. If these wastes are not collected and put to use by manufacturers or not disposed of securely to prevent interaction with the environment, they can pollute and diminish food production and contaminate natural resources in terrestrial and aqueous ecosystems. This may be by rainwater runoff of fecal matter rich in nutrient into food fish habitats (e.g., oceans, estuaries, rivers). The nutrient rich runoff fuels algal blooms that poison habitats (with toxins) or cause depletion of oxygen (eutrophication). As cited previously, both toxins and eutrophication can result in great fish kills. Also, leakage of pollutants in waters from animal husbandry locales into soils can contaminate crops. This might be by unregulated disposal of solid matter that allows rainwater to interact with soils and as previously noted, move dissolved contaminants into soils and by seepage through soils into aquifers thereby contaminating them. As noted in an earlier paragraph, free range stock will not cause such environmental harm when herds are properly tended (e.g., prevent overgrazing) but will have fewer but organically bred food animals to sell.

6.5.2 The Extent of the Manure Waste Problem and Possible Solutions

To appreciate the extent of the animal husbandry waste problem, we can examine the issue with thousands and thousands of tons of manure generated at cattle feedlots. Each head of cattle in a feedlot can drop 15–20 times the mass of manure as does one person. A lactating dairy cow can excrete ~44 times a person's output [8]. Thus, a feedlot with 50,000 cattle would excrete the same mass of manure daily of about 800,000 people. There are 32 such lots licensed in Kansas, USA, state with well over 2,000,000 heads of cattle in feedlots. The sheer mass and volume of manure that has to be disposed of and well cleansed of pathogens and residues of growth hormones (anabolic steroids) and antibiotics used in animal husbandry. This can be accomplished in three ways. Done properly, this can reduce the risk of contamination of water and soils that can negatively affect food delivered to urban centers as well as clean water security, and limit the evolution of antibiotic resistant strains of bacteria that threaten the healthcare community, topics that will be treated in following paragraphs. Pigs and sheep raised in close quarters and that are infused with growth hormones and antibiotics generate masses of solid wastes that should be treated as described below for cattle.

An initial disposal method used was to spread raw manure over acreage with its contained pathogens and residues of growth hormones and antibiotics. Mobilization of these pollutants by rainwater, melting snow, or floodwater has contaminated nearby surface waters by inflow, and via leakage into soils and seepage through soils into aquifers. Ingestion of these residues from soil grown and irrigated foodstuffs that are not thoroughly washed can be health threatening over time to consumers (by bioaccumulation in body organs). In the end, the volume of manure generated daily overwhelmed the available acreage and required the use of other disposal methods that would not endanger people and ecosystems. Planning teams for building a new city have to determine possible health risks may originate from previous use of what appear to be good terrain for development but that may have been for animal husbandry If so, the teams have to assess how and where wastes were disposed of and the efficiency of such disposal.

One disposal method uses heat generated by manure fermentation to kill pathogens at or close to the waste accumulation site. Masses of manure are stacked up and with time ferment and heat up from within to temperatures that eliminate pathogens, growth hormones, and antibiotics (>131 °F but <160 °F for 15 days). Higher temperatures will kill beneficial microbes. A problem with this method is that the high temperatures do not reach the outer edges of the stacks leaving a significant amount of the contaminants active [9, 10]. These would have to be retreated.

Another fermentation method to cleanse cattle manure is to compost it in large rotatable drums that hold thousands of pounds of manure that are turned once a week for six weeks. This mixes and aerates the manure thus distributing the built up heat (>131 °F but <160 °F) throughout the mass eliminating >90% of the health threatening components. In this way, the tonnage of manure changed to compost is reduced by ~60%. The sterilized nutrient rich manure product can be dried, bagged, and sold as fertilizer that also enhances moisture retention by soils. The manure product can also be dried, pressed into briquets and combusted to produce energy [9, 10]. This can be an economic benefit to feedlot owners.

Chickens produce an average of 0.054 pound of manure daily [8]. A commercial operation with one million chickens would produce 27 lbs. (12 1/4 kg) of manure daily or 54 lbs. (24.5 kg) daily for two million chickens. The manure is nutrient rich (nitrogen and phosphorus) but is a carrier of disease and internal parasites. When this is spread on fields as a disposal method, runoff during rainstorms can carry the waste to discharge into productive rivers, estuaries, and oceans that may be serving the interests of urban populations. As discussed earlier, the nutrient rich matter stimulates algal and toxic dinoflagellates growth. When algae die, they decompose using oxygen in the water to the point where there is no oxygen for fish and a resulting massive fish kill including oysters and clams. When toxic dinoflagellates die, they release a neurotoxin into the water that manifests itself in the ocean as a "red tide" and millions of dead fish. This causes a break in the aqueous food web and temporarily diminishes food fish security for coastal communities (cities and towns, villages) as well as causing a temporary economic disruption for fisheries. Instead of spreading the chicken manure on fields and causing ecological and economic problems, many commercial poultry producers are composting and aging the manure for 2–3 months and then recycling it as fertilizers. The Perdue chicken producer set up a large scale controlled composting operation that has profitably processed 650,000 tons (591,000 tonnes) of chicken manure (waste from more than 48 million chickens) during 16 years to produce certified organic fertilizer [11].

6.5.3 The Urine Waste Problem

Urine from feedlots that is not captured and properly treated before discharging it at the surface can seep through soil into groundwater and contaminate it with the chemical nitrate. Groundwater carries the nitrate slowly down flow commonly at 1–5 ft. (30–150 cm) a day depending on the aquifer rock permeability. If there is a population 5 mi (8 km) down flow ("out of sight, out of mind") that uses the groundwater as its primary source, a 5 ft. a day flow rate would bring the nitrate contaminated water from a feedlot to unaware users in about 14 years. If urine input were to stop, it would take 14 years for the contaminated aquifer water to clear a municipal or individual well 5 mi away. If the aquifer flow rate is less than the 5 ft. (150 cm) a day, then the time to reach a user or for contaminated water to clean the well is longer. This is a possible problem planners have to deal with. Untreated nitrate-laced water can be unknowingly used to prepare baby formula and the subsequent bioaccumulation of nitrate in a baby's system has caused "blue baby sickness", a type of nitrogen narcosis. It is important for a planning team that proposes to use aquifer water for a new city or more likely redevelopment of a city to accommodate its growing population to do four things. First is to have analyses made of the water to check its potability or need for treatment. Second is to determine what projects are or were up flow and what contaminants were or could have been released as untreated effluents onto aquifer recharge areas. Remember that recharge areas are extensive so that surveys of what possible contaminant sources might be in the recharge zones or have been there in the recent past have to be broad and fairly far reaching. Third is to determine the flow rate of the aquifer water. Fourth, if treatment is necessary, planning on how to treat the water at a well site or at the tap and the cost of doing so determined. This precludes an assessment of whether a collection, treatment facility and clean water distribution network is feasible given the costs for its construction, operation, and maintenance.

A chemical element that has poisoned groundwater, sickened many, and put millions in India and Bangladesh at health risk before its presence was discovered is arsenic (As). The arsenic has nothing to do with animal husbandry but is worth mentioning here to emphasize the fact that ground water has to be analyzed at its source and along its flow path. The arsenic originated from excessive seasonal extraction of groundwater for irrigation from 20,000 tube wells during the "green" revolution. The water was used as well for drinking, cooking, and personal hygiene. Excessive pumping lowered the water table during a crop growing season and exposed the arsenic bearing mineral pyrite in the aquifer rock to oxidation that released an arsenic toxin to groundwater when recharge took place after a harvest. Once the health problems were diagnosed and related to arsenic ingestion, methods were designed to remove this toxic metal from the groundwater [12].

6.6 Threats to Public Health from Growth Hormones (Anabolic Steroids) and Antibiotics Used in Animal Husbandry

For many of the ~4 billion people living in cities/mega-cities in 2018 or of the 6.9 billion projected to inhabit them in 32 years (2050) in developed, developing, and less developed countries, meat from ruminants, foul, or fish is or will be a common food in their diets. This often comes from distant sources. Methods of production of these comestibles varies according to the standards established by health agencies charged with protecting the food supplies and security for their citizens. For some, this means disallowing the use or overuse of chemicals and/or vaccines to stimulate animal growth or as protection against diseases especially in large scale operations such as cattle feedlots, dairy farms, pig pens, commercial poultry farms, and aquaculture. This applies as well to smaller meat production set ups. Planning teams should consider food sources/security when they assess the availability of food needed to sustain people that will inhabit a purpose built city or existing cities with rapidly growing urban populations. These are discussed in the following sections.

6.6.1 Growth Hormones Residues in Meat Products

As important as dealing with solid and liquid wastes generated in food supply businesses is dealing with the residues from growth hormones (anabolic steroids) implanted in cattle as they enter feedlots in the United States. There are three natural and three synthetic steroids in use. At slaughter houses, these are found in measurable concentrations in beef (muscle and fat), and organ meats (e.g., liver and kidneys). These may be consumed by most urban citizens that do not opt to buy organic meat products, most often because of higher cost. In addition to beef cattle, the steroids are also allowed for use in dairy cows and sheep, but not for pigs or poultry. The use of the recombinant bovine growth hormone (rbGH) was banned in meat production in the European Union in 1988, and subsequently in Canada, Australia, New Zealand, and Japan because the rbGH gives rise in milk to IGF-1, a chemical that helps some types of cells to grow and may influence the risk of prostate, breast, or colorectal cancer [13, 14]. Because the growth hormones cause unnatural cell growth, a phase in the development of cancer, it is incumbent on governments to further investigate banning their use in animal husbandry. Also, their ingestion by humans and possible bioaccumulation in vital organs may alter fetal development and bring about early puberty. Early puberty in young girls and breast development could be a target for unnatural cell growth and a catalyst for breast cancer [15]. There is a paucity of research reports on steroid sex hormone residues and meat consumption. However, a 2007 paper reported on meat consumption during pregnancy associated with sperm concentration and volume in sons of women in two groups: low meat consumption and high meat consumption. The sperm concentrations and volumes were 24.3% higher for the low meat consumption group, but for the sons of the high meat consumption group, 18% had sperm concentrations below the WHO standards threshold for sub-fertility. This was three times greater than the sons of the low meat consumption group [16]. The probable health issues that arise from the secondary ingestion of growth hormones should drive the United States Congress and regulators from other governments to increase financial support in this research field.

6.6.2 *Antibiotics Residues in Meats*

Equally important to human health in cities and villages as regards animal husbandry as well as in food producing ecosystems is the use of antibiotics in feed for cattle and other food animals raised in feedlots or other close quarters (e.g., pens, cages) and the sometimes poor hygienic condition there. In 2009 in the United States, 28.8 million lbs. (13.1 million kg) of drugs were sold for livestock use and 7.2 million lbs. (3.3 million kg) for human use, ~80% vs. ~20%, respectively. Many of those antibiotics that are commonly used to treat human bacterial infections were sold and used for livestock [17, 18]. The drugs are employed to stimulate livestock growth and to protect the enclosed animals from sickness. Medical professionals fear that the sheer volume of the antibiotics being used abets the mutation of bacteria to antibiotic resistant strains. Before mutation, antibiotics could control most pathogens that cause sickness in humans. Medical groups have suggested that antibiotic use should be limited to feedlot cattle that are sick and not to every head in the feedlot. This would spectacularly reduce the volume of antibiotics used and slow the evolution of mutated strains of antibiotic resistant bacteria. As of January 1, 2017, farmers in the United States needed a prescription or Veterinary Food Directive to use antibiotics that are important to humans. Drug resistant bacteria pose a public health threat that is not a major focus of pharmaceutical research into effective antibiotics against the resistant bacterial strains. This is likely because pharmaceutical companies are tasked to deliver maximum dividends to their stockholders. Thus, there is no heavy investment in antibiotic research that is not foreseen as yielding strong financial returns. However, to counter this, government funding in the United States and other countries is being directed to government and university research laboratories to focus on the development of drugs to be effective against the resistant bacterial strains.

6.7 Aquaculture

Fin fish, mollusks, crustaceans, frogs and other aquatic animals and aquatic plants are raised in aquaculture operations. Fish is a major food/protein source for people in city and rural setting and especially for people that live in coastal communities proximate to oceans and lakes in many countries. The mass of food fish raised in aquaculture systems has been steadily increasing as ocean wild fish capture has decreased because of overfishing and enforced global legislation that reduces capture allotments in order to allow ocean fish stocks to replenish. Aquaculture now produces 50% of all seafood for human consumption [19]. By 2030, the aquaculture production is projected to reach 62% of all food fish [20]. As such it is important for planning to ensure first that food security from this source can be sustained to help feed annually increasing city and global populations and second that aquaculture products are safe to consume. Aquaculture has similar problems as agriculture when growth hormones (anabolic steroids), antibiotics, or in this case water treatment chemicals are used in the operation. Also, care must be taken to assure that the chemicals and fish feces from containerized pools (tanks) that constantly build up are regularly removed and treated before subsequent discharge into the environment. The dirty waters, if discharged untreated into waterways may be detrimental to the safe human use of water from them and may also affect life forms there by causing mutations or by killing them. In well planned special projects the dirty water may be used as part of a looped project.

6.7.1 A Looped Project in Eritrea

An example of a looped project in the Eritrean coastal desert begins with a 5 km (3 mi) long canal that brings Red Sea water into concrete lined shrimp tanks to prevent leakage into surface ecosystems or the underlying aquifer. The shrimp species (*Penaeus vannamel*) and Indian prawn (*Penaeus indicus*) tolerate changes in salinity temperature, pH, and oxygen. The sea water from the tanks is then channeled into three salt lakes for aquaculture cultivation of Tilapia and milk fish (*Chanos chanos*) for food and to make products including shrimp feed from recycled fish heads. Waste from the lakes, rich in organic nutrients, is carried by water to fertilize *Salicornia* farming. *Salicornia* is a succulent plant that can grow normally in fields irrigated with salt water. Its seeds are pressed to give oil and young shoots of *Salicornia* are exported to Europe where they are considered a gourmet vegetable. The waste from *Salicornia* is used to make particle board, firebrick, and animal fodder. The seawater then seeps into the soil and flows back to the Red Sea. Although the looped process worked well, the business was not economically profitable because of management problems and lack of enough production to meet demand. Development of sustainable aquaculture in coastal Eritrea and other countries can contribute significantly to urban centers protein rich food supplies and is ongoing [21].

6.7.2 Aquaculture in Ocean/Estuarine Waters

Net-penned fish aquaculture (plus shellfish [shrimp, oysters], algae) in estuarine and nearshore ocean environments that allows free water exchange with the pens has a lesser ecosystem waste problem than the contained pseudo ecosystems. However, if the fish have been genetically modified to stimulate growth and prevent disease, there is the problem of escape from a pen and interaction with wild species that may negatively alter the wild species genome.

Aquaculture products should be checked periodically by certified consultants to detect any unhealthy contents of chemicals in their flesh in the same way ocean food fish (e.g., tuna, swordfish, tilefish, king mackerel, shark) are tested for mercury (a neurotoxin). Results would then be subject to recommendations made by the medical community for rate of consumption as is the case for mercury in the species named above.

References

1. Chang, Q., Wang, W., Regev-Yochay, G., Lipsitch, M., & Hanage, W. P. (2015). Antibiotics in agriculture and the risk to human health: How worried should we be? *Evolutionary Applications, 8*, 240–247. https://doi.org/10.1111/eva.12185.
2. HM Government and Wellcome Trust. (2014). O'Neill, J., Chair. Review on antimicrobial resistance. .Antimicrobial resistance: Tackling a crisis for the health and wealth of nations. 15 p.
3. Katrime Integrated Health. (2016). The role of human health and animal health in antimicrobial resistance. National Collaborating Centre for Infectious Diseases, 13 p. Online www.nccid.ca/publications/role-human-health-animal-health-antimicrobial-resistance
4. Zdziarski, I. M., Edwards, J. W., Carman, J. A., & Haynes, J. I. (2014). GM crops and the rat digestive tract: A critical review. *Environment International, 73*, 423–433. https://doi.org/10.1016/j.envint.2014.08.018.
5. Tabashnik, B., & Carriere, Y. (2017). Insect resistance to transgenic crops: Second decade surge and future prospects. *Nature Biotechnology, 35*, 926. https://doi.org/10.1038/nbt.3974.
6. University of Arizona. (2017). Pest resistance to biotech crops surging. Summary of above reference. In Science Daily, Oct. 10. Unpaginated.
7. Shaw, J. (2018). A new green revolution? *Harvard Magazine*, March–April, pp. 44–48.
8. Fleming, R., & Ford, M. (2001). Humans versus animals – Comparison of waste properties. 4 p. Online. www.ridgetown.com/research/documents/f;eming_huvsanim0107.PDF
9. Alberta, Agriculture and Forestry. (2005). Manure composting manual. Unpaginated, Online. www1.agric.gov.ca/%24department/deptdocs.nsf/all/agdex8875
10. Augustin, C., & Rahman, S. (2010). Composting animal manures: A guide to the process and management of animal manure compost. North Dakota State University, 8 p. Online. www.ag.ndsu.edu/manure/documents/nm/478.pdf
11. The Daily Record. (2017, November 9). Perdue farms now selling recycled poultry manure as fertilizer. Online. After Associated Press, 2011, May 10. thedailyrecord.com/2011/05/10/perdue-farms-now-selling-recycled-poultry-manure-as-fertilizer/
12. Nickson, R. T., McArthur, J. M., Ravencroft, P., Burgess, W. G., & Ahmed, K. M. (2000). Mechanism of arsenic release to groundwater, Bangladesh and West Bengal. *Applied Geochemistry, 15*, 403–413.

13. American Public Health Association. (2009). Opposition to the use of hormone growth promoters in beef and dairy cattle production. Policy No. 20098.
14. American Cancer Society. (2014). Recombinant bovine growth hormone. Online. www.cancer.org/cancer/cancer-causes/recombinant-bovine-growth-hormone.html
15. Aksglaede, L., Sorensen, K., Petersen, J. H., Shakkebaek, N. E., & Juul, A. (2009). Recent decline in age at breast development: The Copenhagen puberty study. *Pediatrics, 123*, e933–e939.
16. Swan, S. H., Liu, F., Overstreet, J. W., Brazil, C., & Shakkebaek, N. E. (2007). Semen quality of fertile US males in relation to their mothers' beef consumption during pregnancy. *Human Reproduction, 22*, 1497–1502. https://doi.org/10.1093/humrep/dem068.
17. Hollis, A., & Ahmed, Z. (2013). Preserving antibiotics rationally. *New England Journal of Medicine, 369*, 2474–2476.
18. Katrime Integrated Health. (2016). *The role of human health and animal health in antimicrobial resistance*. Winnipeg, MN: National Collaborating Centre for Infectious Diseases 13 p.
19. NOAA. (2012) Fisheries. Online. www.nmfs.noaa.gov/aquaculture/faqs/faq_aq_101.html
20. World Bank. (2013). Fish to 2030 – Prospects for fisheries and aquaculture. World Bank report no. 83177-GLB, 80 p.
21. Habte-Tsion, H. M. (2016). Sustainable aquaculture development and its role in the food security and economic growth in Eritrea: Trends and prospects. Presented at the International Conference on Eritrean Studies, July 21, Asmara, Eritrea. 9 Sections. Online. www.eritreaembassy-japan.org/data/Aquaculture-Development.pdf

Chapter 7
Master Plans to Meet Basic Public Health Needs for Projected Growth in Urban Populations

7.1 Introduction to Planning Challenges

As noted at the beginning of this book, demographers project that the global urban population will grow from ~4 billion people (of 7.5+ billion) in 2018 to 6.9 billion (of 9.8 billion) by 2050. Most of this demographic change will take place in Asia and Africa as countries focus on economic growth through industrial development and employment opportunities that industry and its service businesses (suppliers) create for citizens. Because of jobs open in construction for industrial development, (e.g. infrastructure), in industries as they go into production, presumed access to education for children, and better health care, workers flocked to and are still flocking to urban centers. Unfortunately many growing cities, and mega-cities that grew from them, did not adequately plan on providing quality shelter for workers serviced by basic requirements as discussed earlier in this text. These include safe water, access to adequate sanitation, electricity, gas, and efficient transportation to move workers (if they are employed) to their jobs and back to their homes. Nor did all municipalities do thorough environmental impact studies. For example, one devastating result is that there has been air pollution from gases, aerosols, and fine size particles (<2.5 μ) emitted by industries. Globally there were 4.2 million premature deaths abetted by air pollution in 2015. In China alone, 1.11 million people died prematurely abetted by inhalation of fine size particles with India was not far behind with 1.09 million premature deaths abetted by fine size particle inhalation [1]. There was damage to agricultural soils by acid rain thus threatening internal food security. Instead, governments exhibited tunnel vision on industrial advancement but not on caring effectively for their growing populations. Only recently, as a result of citizen protests (e.g., of death and sickness by smog, of Pb (lead) poisoning of children from residentially located factories) and international condemnation of problems such as just cited, have governments (e.g., China) begun to respond to environmental dangers with legislation that is being enforced albeit ofttimes spottily.

F. R. Siegel, *Cities and Mega-Cities*, SpringerBriefs in Geography,
https://doi.org/10.1007/978-3-319-93166-1_7

7.2 What Should Be Done? What Can Be Done?

It is easy to write what should be done to improve health conditions and quality of life necessities for cities where these are lacking for important segments of the populations that have immigrated to them in large numbers in relatively short periods of time. This gives rise to overcrowding, unemployment, social stress, and overworked infrastructure that doesn't reach all. The increased population can cause an overuse of water supplies, pollution of air, water, and soil, and other factors that harm the urban citizenry quality of life.

Conversely, it is extremely difficult to prioritize what can be done given economic restraints on rehabilitation funds that are available and the political factions lobbying for their interests. And then there is the fact that corruption exists. Without eliminating or greatly curbing this corrupt scourge of society, there will be much less funding to rehabilitate problems than funds that are granted. These barriers notwithstanding, we can set prioritization for funding projects. A listing of environmental and quality of life problems from some of the densely populated mega-cities establishes communality that can assist in deciding where to put redevelopment funds so that ultimately there will be meaningful positive benefits for all urban citizens.

The communality of human and environmental issues of mega-cities in Asia and Africa provides large scale examples of these issues that can affect other less populated cities. The issues arise from rapid unplanned or mismanaged urbanization and random, unregulated or regulated but not enforced, manufacturing and industrial development. Many such issues that are considered in following paragraphs affects citizens in a city differently depending on economic status but have a greater negative impact in agglomerated, generally least economically-advantaged populations in peri-urban areas.

7.3 Communality of Problems in Major Urban Populations

The pubic health, living conditions, and environmental issues that confront urban populations worldwide (especially in Asia and Africa and to some degree in South America), are the result of two principal factors. These are first the rush of governments to industrialization, and second, population growth and demographic changes that governments did not prepare for. Industrialization creates employment in multiple sectors and this attracts rural citizens to move where there are employment opportunities. The rural to urban population transfer is further driven as already noted by better access for families to healthcare clinics and education for workers' children. Common problems are evident where there has been funding reduced by corrupt practices, and misdirected planning for use of available funds.

I have reviewed the human-related and environmental-related problems associated with Delhi, Mumbai, and Kolkata in India, Karachi and Lahore in Pakistan,

Table 7.1 Cited mega-city population build-up 2008 and 2016 and annual growth

Mega-City	2008 Population		2016 Population	Annual Increase
		Census		
Delhi	18,000,000	(2001)	26,495,000	566,300
Mumbai	19,530,000	(2001)	22,885,000	~221,600
Kolkata	15,010,000	(2001)	14,950,000	~115,500
Karachi	9,380,000	(2005)	23,545,000	~1,416,500
Lahore	6,810,000	(2005)	10,665,000	~385,500
Dhaka	7,310,000	(2001)	16,820,000	~634,000
Manila	19,550,000	(2007)	24,245,000	~586,800
Lagos	8,860,000	(2006)	13,360,000	~500,000
Kinshasa	8,190,000	(2007)	11,855,000	458,100
Lima	7,750,000	(2005)	11,150,000	~340,000
Cairo	16,750,000	(2006)	16,225,000	(−58,300)

Census data for 2008 listing in parentheses for annual increase number. 2008 is the start of global economic crisis that likely pushed migration to urban centers [2, 3].

Dhaka in Bangladesh, Manila in the Philippines, Kinshasa in the Democratic Republic of Congo, and Lagos in Nigeria. The commonalities and reasons for problems that are detrimental to public health, to improvement of living conditions for citizens, and to environments (ecosystems) that provide resources for inhabitants of these mega-cities are a matter of record as reported by historically reliable sources. These are discussed below.

7.3.1 Urban Population Growth: Examples in Mega-Cities

As noted in previous sections, urbanization is driven principally by industrial development that has attracted the great influx of populations to cities and with it many of the human and environmental/ecosystem issues suffered increasingly in recent years. This will intensify in the near future (e.g., in 2050) if demographic predictions that city populations worldwide will grow by 2.9 billion people from the 2018 total of ~4 billion. Table 7.1 illustrates the magnitude of annual growth in mega-cities named in the previous paragraph. Mega-cities other than those given in Table 7.1, and large cities with lesser populations are coping with the same issues. It should be noted that the population figures given in Table 7.1 represent census figures provided by national governments and may not be truly accurate [2, 3]. However, the orders of magnitude are reasonable and give a good sense of the rate of population growth.

For Delhi, the rate of growth from 2001 to 2016 was approximately 3%. For Karachi, the rate of growth from 2005 to 2016 was the fastest at approximately 15%. Unlike most mega-cities, Cairo showed a loss of population from 2008 to 2016 implying an minor exodus of about 0.3% annually to locations that offered better opportunities and/or better personal security. The human issues arise when a municipality can not provide for the burgeoning population growth physically, biologically, economically, or functionally because of lack of or misdirected urban planning as to spacing for residences, industries, businesses, and government offices, versus well thought out spacing described earlier in this text. The environmental (health) issues arise from unfettered industrial development without regulations that can protect citizens and irregular infrastructure activity. This results in pollution and other ills such as health threatening poorly accessible or inaccessible clean water as total water supplies diminish lessening per capita availability. The lack of sewered sanitation systems and solid waste collection and disposal add to environmental problems including inattention to workplace safety.

7.3.2 Air Pollution

The communality that brings on health problems, population sustainability issues, and environmental problems and their causes may have solutions imposed and enforced by legislation. Others have economically viable solutions or solutions that are too expensive for a country to afford but some of which can be implemented for mitigation with available economic resources.

Air pollution is a common issue that has caused respiratory problems (e.g., to some degree in 70% of the population in Kolkata). These likely originate from the inhalation of fine-size (<2.5 μ) particulates and toxic fumes that are emitted mainly from industries, especially those in residential areas (e.g., in Lahore) or in populations downwind of them. Added to this are particulates and gases from poorly serviced, unclean internal combustion vehicle engines with excessive exhaust. Fine size particulates are also released to the atmosphere from the use of generators where electricity does not reach (e.g. peri-urban populations), from road and construction dust and brick kilns (e.g., Dhaka), as wind-borne particulates from arid areas (e.g., Cairo), from coal ash, and from open-air waste burning (e.g., Cairo, Kinshasa, Lagos). Toxic chemicals from industrial processes add to the air pollutant loading by gas emissions and aerosols with heavy metal pollutants. Added population means increased vehicular traffic, more consumption and hence more domestic/commercial wastes to be disposed of as well as sanitation wastes, some by burning. This can add to a city's air pollution condition if there are no municipal actions (physical and/or legal) to reduce such pollution sources. For example, and as noted previously in the text, this is the case where crop residue burning in India adds to air pollution loading in Indian cities. The same action poisons the air in Lahore, Pakistan. This emphasizes the fact that planning is necessary, now and for the future to diminish the air pollution and other problems that can and often do plague cities whether mega-size or smaller. This is especially true where industrial development plays a large role in a governmental master plan. This will be discussed in later paragraphs.

7.3.2.1 Possible Solutions to Urban Air Pollution

There are three previously cited possible solutions to an urban air pollution problem caused by industrial operations. One is to close down the polluting industries of which there may be many in residential/commercial areas and/or on the outskirts of and upwind of cities. This is questionable because it will put breadwinners out of work hurting families, and reducing municipal tax revenues from workers' salaries, taxes they pay for goods and services, and taxes paid by industries themselves. As cautioned in an earlier Sect. 3.2.1.3, if closure is done, then it is necessary to make sure that toxic materials are not left behind in an abandoned structure, in surface or buried storage facilities. It is also important to know where industries disposed of toxic wastes in the past and disposal methods (e.g., buried raw or in containers, their depths) in an assessment of terrain use where there could be uncatalogued disposal sites.

A more feasible solution that avoids closure of industries outside a city proper and the ramifications given above, is to install up to date best available technology to capture pollutants. Pollutants produced vary greatly with an industrial product so that pollution capture/control systems are tailored to remove a particular industry's specific pollutants generated during production processes.

These might include fine size (<2.5 μ) and larger particulates, fly ash (coal ash), and chemicals, as gases or aerosols, such as sulfur dioxide, hydrogen chloride, mercury, volatile organic compounds, dioxins, odiferous hydrogen sulfide, and other toxic fumes as well as whatever additional toxins an industry emits. This presents an economically viable alternative to closing down factories not in cities themselves. Pollutant capture/control equipment should be constructed in modules so as to be able to readily adapt to technological advances that reasonably increase pollution capture efficiency.

Relocation of industries to industrial parks/zones a good distance away and downwind/down flow from residential and commercial areas is feasible. This depends on the profitability of industries to economically support a move and municipality incentives to keep an employer. Incentives can include a municipality's economic contribution to extend required infrastructure (e.g., transportation for workers, railroad spurs) if it is not already in place, and/or a tax holiday or reduced tax rates for a reasonable period of time. Industries relocate from one country to another for these reasons but more so for lower salaries to workers and perhaps lax or unenforced environmental regulations both of which increased profits but ultimately can hurt workers, their families, and all urban citizenry in a host country. Smaller industries are less likely to move overseas because of the cost of doing so and estimated economic benefits balanced against incentives to move to an industrial park such as noted above.

It would seem that a combination of moving polluting industries out of residential areas or from upwind of urban populations and legislation that requires use of best available pollution control equipment and its maintenance and use is a viable option. However, this is possible if **long-term** benefits/costs ratios are sufficiently positive economically. In the case of a mega-city like Delhi, where only about 20%

of industries are in industrial parks with 80% scattered through commercial and residential areas, a relocation is not realistic economically in the foreseeable future. The aim then would be to mitigate Delhi's air pollution problem by legislating and enforcing the installation, maintenance, and use of pollution control equipment in addition to putting an end to crop residue open air burning. Will the Indian government give tax benefits to support this or even to some degree subsidize this? Are the premature deaths of 1.09 million people in India during 2015 (<0.1% of the population) from air pollution enough to loosen the country's industrial sector purse strings? Other cities such as Lahore also have industries and populations that suffer from air pollution they emit in residential areas. In a report on ambient air pollution, the World Health Organization presents a global assessment of exposure and burden of disease [4].

7.3.3 *Lack of Safe Water*

In-migration from rural areas, natural population growth, and in recent years normal immigration from other nations plus refugees seeking asylum from war is a communality situation that has been overwhelming for many city/national governments. This includes a lack of development of complementary infrastructure that can deliver clean/safe water to all residents. In such a case, the demand for clean water exceeds the supply. The use of unclean water has led to and continues to lead to public health ills that endangers peoples' well being and lives of the very young and the aged. WHO reports that in 2012, 1.5 million people died of water borne diseases of which 842,000 were attributed to lack of safe drinking water, inadequate sanitation, and hygiene, Children under 5 years old accounted for 622,000 deaths of the deaths, 361,000 of which could have been avoided by hand washing with safe water [5]. One-sixth of all deaths in Manila are from water-borne diseases and poor sanitation (e.g., diarrhea, intestinal worms, cholera, and typhoid [for which there is a vaccine]).

The cities listed in Table 7.1 suffer from not being able to meet the clean water needs of millions in their edge populations. The reasons for people having to use polluted waters begin mainly with a city's lack of infrastructure networks that bring clean water to and access to adequate sanitation to large numbers of under serviced populations within a city or to peripheral informal settlements. Environmental unawareness by people that practice open defecation and uncontrolled waste disposal complicates the clean water and public health problems.

The origin of pollutants that make available water undrinkable/unsafe fall into several categories that highlight the communality of sources in several of the megacities listed in Table 7.1. These extend to other highly populated cities in Asia and Africa. The water problems result in part from colonial powers that established good water service for core cities but left no plan that was followed or developed by succeeding national governments that assumed power after the colonialists exited. The problems result in part also because of lack of maintenance by existing municipal/national governments of the water collection, treatment, and distribution network. These include repair or replacement of broken pipes that lose water cleansed at treatment plants such as in Lagos. Here the state utility company estimates at least 1/4 of the treated water supply is lost to leakage with a total unaccounted for loss of 60%, probably from people breaking into pipes to get clean water. In Cairo, the utility company estimates the 35–50% of treated water is lost to leakage. Add to this, sanitation systems with sewage pipes that cracked with age and lack of maintenance and now leak human wastes into the subsurface, polluting groundwater resources.

7.3.4 Lack of Access to Adequate Sanitation

The lack of sanitation facilities and of environmental health education of the citizens in some major cities has allowed a continuation of the practice of open defecation that poisons waterways directly or via transport by rain runoff. Leakage from poorly maintained septic tanks contributes to water contamination by sewage. Add to this, lack of collection of garbage and other solid wastes or their collection and deposition at non-engineered dumps so that lets rain leach through the wastes mobilizing health-threatening toxins that can seep into aquifers. Where wastes have been piled high onto un-engineered dumps, there have been waste slides that have killed people searching through the dumps for sellable refuse (e.g., glass bottles metals, electronic pieces). In March 2017, outside Addis Ababa, Ethiopia, a slide from a massive garbage dump killed 46 people with many people missing. This is not a unique happening as 143 people were killed by a waste slide near Bandung, Indonesia in 2005, and 50 people were killed in 2008 by a waste slide outside Guatemala City. Municipalities and their planning consultants have to consider this potential dump-slide hazard in redevelopment proposals that are purposed to improve the safety of their less advantaged citizens that scour dumps by giving them another opportunity to earn money. This maybe by establishing collection and recycling centers that can employ these citizens doing the same waste picking but under safe conditions and at a salary determined by the income from sales to recycling businesses. Industrial wastes disposed of at un-engineered dump sites contribute to the contamination of surface and aquifer water supplies with potentially toxic heavy metals (e.g., As, Pb, Ni, Cd, Cr, Hg). If ingested by citizens and bioaccumulated in human organs over a period of time, one or more of these toxins can cause illness or death. Contamination by wastes from markets and slaughterhouses is an important cause of water contamination and epidemic of intestinal diseases in Manila.

7.4 Mitigation of Water Problems

Pollution makes water undrinkable at the threat of sickness mostly for poor people living in shantytowns/slums with their growing populations in many major cities and mega-cities and lack of access to clean water. It also affects rural towns and villages that suffer from insufficient supplies of safe water. This can be greatly mitigated if not entirely eliminated as a doable but long term project. Correcting the problem is done first via investment in education that stresses environmental awareness about disposal of human wastes and garbage and its impact on societal health. Second, is a governmental commitment to support the environmental awareness programs, not with words alone but with actions including investment in sanitation and waste collection and secure disposal, and waste water collection, treatment, and clean water distribution networks. Third, mitigation or elimination of sources of water pollution mainly from other than domestic/commercial sources can be accomplished by legislation that requires cleansing of manufacturing/industrial and agricultural effluents to drinkable standards, safe disposal of solid wastes from these sources that might otherwise invade water sources, and enforcement that punishes cheaters with financial penalties, shut down of operations, or incarceration.

7.4.1 Awareness Through Education and Engineering Advice

Education can bring environmental awareness to citizens by explaining consequences that they can suffer from their actions. This is best illustrated with spectacular results in Bangladesh. In 2003, more than 60 million people of the ~138.5 million Bangladesh population (43%) practiced open defecation. This caused a high incidence of water-borne diseases such as diarrhea, dysentery, cholera, and typhoid as bacteria moved from feces into water resources. National productivity was hurt by this because of worker days/years lost to sickness. The Bangladeshi government invested 25% of the national development budget over the next 12–13 years to educate its citizens on how improved sanitation habits could reduce illnesses and prevent deaths of small children. Engineers worked with citizens showing them on site how to build and maintain latrines. This was done nationally by public health educators and government engineers, carefully and thoroughly with purpose and without haste. The result was that the 43% of the country's population cited above that had practiced open defecation dropped to 1% (~1.4 million people) in little more than a decade. Cities with problems of human wastes disposal problems that poison water supplies such as Kinshasa and Lagos in Africa, should consider adopting the Bangladeshi approach to their public health and sanitary engineering capabilities. India has a more severe problem with 48% of its citizens, close to 600 million people, who practice open defecation or 10 times the number in Bangladesh in 2003. Major cities such as Mumbai, Delhi, and Kolkata suffer from this problem of the spread of water-borne diseases from water contamination by human feces.

Prime Minister Modi recognizes the severity of the problem but prefers to attack it with the Smart Cities Project described earlier with an initial budget of US$15.4 billion. This attacks the open defecation problem in pilot projects for a small percentage of but not all of India's at risk population by stressing the installation and use of toilets or latrines or daily collection and safe disposal of honey pot wastes. This is Mr. Modi's approach rather than using the proven and successful Bangladeshi approach by investing in countrywide education by public health professionals and sanitary engineers. These specialists are available and prepared in India and need only investment funds and time to begin reducing the country's open defecation problem. This could surely be a complementary approach to the Smart Cities Project that can readily be put in place and hasten India's attack on the practice of open defecation and the water-borne diseases millions of citizens suffer from.

A second environmental awareness lesson depends partially on citizens but mainly on municipal governments. This is the regular collection of garbage, trash, and other domestic/commercial wastes for disposal. At present, in some major cities and mega-cities, this waste is dumped by citizens, unserviced by regular waste collection, in gutters, in canals, and on green spaces that have water running through them, thus contaminating down flow water for human use. Secure disposal may be in engineered dumps that prevent leaching of rainwater into subsurface aquifers as may exist by using impermeable linings with a fall slanted towards a collection tank from which leachate can be recovered and properly treated before use or disposal. In Western Europe, Rome sends train car loads of waste to Austria where much is separated and becomes fuel for an electricity generating plant. This would seem to be a perfect investment for African nations and India where there is not enough electricity generated to serve all citizenry 24/7. Once the human waste problem and that of regular collection and secure disposal of garbage and other domestic/commercial garbage and trash is resolved, two major sources of water pollution will have been eliminated or greatly reduced.

7.5 Clean Water Projects and Funding

Projects that eliminate or greatly reduce sources of water pollution require funding. In addition to self financing by municipalities and national governments themselves, there are other sources of financial help to carry these projects to a positive conclusion. These may be as loans from the World Bank, from regional development banks (e.g., Asian, African, Inter-American institutions), grants from rich nations such as those in the EU, Scandinavia, the USA, Canada, Japan, and others with substantial foreign reserves (e.g., China). However, these loans and grants will require as a condition for a loan or grant rigidly enforced transparency that precludes bribery or extortion to obtain contracts as funds are meted out in parcels. This should be the modus operandi for the stages that follow that will strive to deliver clean water to population that need it. Incarceration and financial penalties should be the norm, not the exception, for those convicted of corruption and those who turn a blind eye to and thus abet this evil practice. No countries are completely free of bribery and extortion but some have a history that accepts corrupt practices as a part of a culture. These appear at the bottom of the Corruption Perception Index prepared by Transparency International [6]. Many cities in those countries where bribery and extortion are considered by some as the norm rather than the exception, such as Kinshasa in the Democratic Republic of Congo or Lagos in Nigeria, have problems with supplying clean water to their citizens and need economic help to accommodate rapid population growth and urbanization. The lending or granting of funds to support projects to bring safe water to populations should be conditional on transparency showing actual project expenses. There should be no funding if this condition is not accepted and a cut off of funds if corruption is revealed during a funding schedule. To reiterate, only when guilty parties are prosecuted, fined, or incarcerated for their evil practices should scheduled funding be resumed. Funding possibilities are discussed in some detail in the following chapter.

References

1. Health Effects Institute. (2017). State of global air/2017. A special report on global exposure to air pollution and its disease burden. Boston, MA. Online. www.stateofglobal air/sites/default/files/SOGA2017_report.pdf.
2. Countries of the World. (2008). The world's most populous cities, metropolitan areas, urban agglomerations. Online. www.allcountries.org/ranks/most_populous_metropolitan_urban_areas_2008.html.
3. Demographia. (2017). World urban areas (built up urban areas or world agglomerations) 13th annual edition, 112 p. Online. demographia.com/db-worldua.pdf
4. World Health Organization. (2016). *Ambient air pollution: A global assessment of exposure and burden of disease*. Geneva: WHO Press 131 p.
5. World Health Organization. (2014). Preventing diarrhoea through better waters, sanitation and hygiene. Geneva, 34 p. Online. http://www.who.int/water_sanitation_health/gbd_poor_water/en/.
6. Wikipedia. (2017). Corruption perception index 2016. See table. Online. www.en.wikipedia.org/wiki/Transparency_International.

Chapter 8
Economic, Human, and Natural, Resources to Support Changes that Improve the Quality of Life for Existing and Future Urban Dwellers

The rehabilitation/redevelopment of urban centers is costly and once begun will take several years to accomplish. It is most efficiently completed in stages by discreet neighborhoods within a city and by prioritized need in adjoining shantytowns/slums. The time to start a renewal process is now as pressures from growing populations increase daily as evidenced by the demographic projection that the 2018 city population of ~4 billion people will rise to 6.9 billion by 2050 fueled as already noted by births, rural citizen in-migration, and sometimes immigration. Today, as previously stated, there are scores of millions of people living in shantytowns/slums in developing and less developed nations especially in Asia and Africa, but also in South America. For most people living in these squalid conditions, basic needs for a reasonably good quality of life are not being provided. What can be expected in terms of social stability as years pass to 2050 and beyond? Restless masses demanding change are a likely answer if indeed municipalities and nations do not show progress to a better life by steadily improving delivery of basic needs to citizens as each year passes. Progress begins slowly with small but real improvements in living conditions in shantytowns/slums by providing basic services that contribute to good public health (safe water, access to adequate sanitation, and regular garbage collection).

8.1 Economic Reality of Urban Rehabilitation to Service Underserved Shantytown/Slum Inhabitants

Efficient use of economic, human, and natural resources are the three fundamentals through which living conditions can be improved for shantytown dwellers living in urban centers or agglomerated to them. By extension, these improvements will benefit the general public as well with overall better public health, greater opportunities for employment, and sustainable essential resources.

F. R. Siegel, *Cities and Mega-Cities*, SpringerBriefs in Geography,
https://doi.org/10.1007/978-3-319-93166-1_8

Constructing a purpose-built major city from scratch is a rarity and requires great investment. The new capital city Brasilia was built in mid-twentieth century from 1956 to 1960 that today houses all national government facilities and 2.8 million people in the city proper and abutting urban districts. The construction took place midst an economic boom by national self financing with the tax income mainly from the mining, agriculture, and forestry sectors. During the 1980s, Abuja was built with national self financing from the oil sector t. During the 1990s, Astana Modern City, Kazakhstan was purpose built, financed mainly by income from the oil sector. It had an initial population of 281,000 people but today has more than one million inhabitants. No reliable costs for the building of the three new capital cities were available. In 2017, estimated costs to build new cities that have twice the population of Brasilia, Abuja, or Astana would likely be in the hundreds of billions of dollars range. In 2017, rehabilitation of cities or creating mega-cities by meshing multiple cities together is extremely costly as described in the following paragraphs.

8.2 Examples of Projected City Rehabilitation/ Redevelopment Costs and those for Purpose Built Cities

8.2.1 India

The costs of rehabilitation/redevelopment of major or minor cities in the twenty-first century are great. They have to be distributed over a relative long term (e.g., 7–10 years) for projects designed to better the living conditions for in city and peripheral neighborhoods with underserved or unserved populations as well as to promote economic development that can provide meaningful employment that includes less advantaged segments of an urban population. This cannot be done all at once but has to be prioritized and completed in stages. For example, Indian Prime Minister Modi put forth the "Smart City" project cited earlier (Sect. 1.5) for planning started in 2015 with US$15.4 billion initially for about 100 cities of various population sizes (US$7.5 billion) and rejuvenation of 500 others with populations greater than 100,000 people (US$7.9 billion) [1]. He expects a final cost for rehabilitated Indian cities to be US$300 billion over a several year time frame. During the initial phase of the program 20 of the 100 cities were chosen to serve as pilot projects with each receiving US$31 million during the initial financial year and US$16 million during each of the following 3 years. As stated previously, priority will be given to solve five basic needs to improve citizens' health and quality of life. As described in section 1.5, these are: (1) to provide safe water 24/7; (2) to provide toilets and daily collection of honeypots from unsewered neighborhoods; (3) improve public transportation including designating one way streets and lanes dedicated to buses or other vehicular transport to reduce traffic snarls; (4) site elementary schools no more that 400 m from homes; and (5) make Wi-Fi available throughout a city [2]. If the designated funding is used prudently and not subject to loss through corrupt practices (bribery, extortion) in the permit-supply-rehabilitation construction stages, there should be positive results within a reasonable time frame. If this initial effort shows positive results for the people in the 20 cities, the way will be paved to invest US$300 billion or more in other Indian cities and rural centers to improve living conditions for all their citizens. This, of course, depends on the continued growth of the Indian economy even as its population continues to grow.

8.2.2 China

China is planning what will be the largest global mega-city. The plan is to merge 9 cities with 42 million inhabitants in the Pearl River delta area, an important manufacturing region that accounts for almost 10% of China's economy. The three cities with the highest populations are Guangzhou, Shenzhen, and Dongguan with 11.7, 8.9, and 6.4 million citizens, respectively. The 9 cities will have a united policy to control industrial pollution. They will be connected by a high speed rail system as part of the 150 major infrastructure projects that will mesh transport, energy, water, and telecommunication systems. The estimated budget is US$196 billion for the rail infrastructure and US$304 billion for the other infrastructure projects and will be self funded [3]. These estimates will likely rise because of overrides and inflation during the optimistic 6 year program to complete the integration of the 9 cities into one huge metropolis that will cover more than 750 mi^2 (1200 km^2). The Chinese government will have to establish checks and balances transparency procedures to prevent or greatly limit corruption that will surely target this undertaking. Under President Xi, the fight against bribery and extortion in China by corrupt officials has successfully begun and will continue. There are relatively few countries with a strong economy and the financial strength to self fund such a high cost undertaking.

8.2.3 Egypt

Egypt has plans for a new capital city for about 5–7 million inhabitants east of Cairo between it and the Red Sea. All national government buildings would move to the new capital city. The expectation is that the new city would relieve the population pressure and congestion experienced in Cairo with its more than 17 million citizens in 2017 but that is projected to double by 2050. The plans are grandiose and call for multiple residential and dedicated districts with 1.1 million homes, multiple business operations, 2000 schools and colleges, 663 healthcare facilities, 1250 mosques and churches with the infrastructure to service all the city's needs, and the creation of one million jobs. The cost to plan and begin to build the initial phase of the new city is estimated at US$45 billion with significant funding to come from outside (e.g., pledges of US$4 billion each from Kuwait, Saudi Arabia, and the UAE). The total estimated cost to complete the new city that will cover an area of 270 mi^2 (432 km^2) when completed in what was hoped to be 5–7 years is US$300 billion [4]. Given that the gross domestic product for the country was estimated at US$342 billion for 2015 and because of serious concerns from potential lenders and investors given the mixed success in past decades in attracting population to new cities in the desert, President al-Sisi deemed it prudent to put a hold on the project.

8.2.4 Nigeria

Abuja is a purpose-built new capital city to replace Lagos as Nigeria's capital.The site selected as a replacement for Lagos' squalor, has a fine climate in central Nigeria, has land for expansion, and is surrounded by hilly terrain and savannah grasslands. The city was completed during 1987 and in 1991 was officially designated as the capital. No cost figures have been published but was likely in the range of US$100s of millions. It has been the fastest growing city in Africa with an annual average population increase of 13.9% between 1991 and 2015 for the metropolitan area. The Abuja City had population at the last census (2006) of well over three quarters of a million inhabitants. The population continues to grow. The increasing population has put a strain on the urban infrastructure so that there had to be a dam expansion and added capacity of water treatment capacity to meet clean water needs. Waste collection capacity also had to be increased. With satellite cities and small settlements, Abuja has a 2017 metropolitan population of almost two and a half million people. As has happened in many urban centers in developing and less developed countries, people in search of employment, family healthcare, and good education for children, flocked in and squatters created slum settlements in the modern city and in the city outskirts. The government did not want Abuja to become another Lagos with its squatter problems. The Center for Housing Rights and Evictions based in Geneva and the Social and Economic Rights Action Center in Nigeria reported that between 2003 and 2007 the Nigerian government razed 31 informal settlements and estimated that 800,000 citizens were displaced and resettled in areas some with poor soils and sometimes without access to enough safe water [5].

8.2.5 Saudi Arabia

In March, 2018, Saudi Arabia announced that a mega-city, by size and not population, would be built in the northwest of the country bordering the Red Sea under the project name NEOM (New Future). It would be designated as a special zone. An area of 26,500 km^2 (10,231 mi^2) is dedicated to the new city and its vast surroundings including 1000 km^2 (625 mi^2) of Egypt's southern Sinai for which Egypt received a Saudi Arabia grant of US$10 billion for planning, and a small area of Jordan. The city is planned as a high tech center that would diversify the oil-based Saudi Arabian economy. Development emphasis would be in several sectors including energy and water, mobility and transport, biotechnology, food, technological and digital sciences, advanced manufacturing, media and media production, and entertainment. The stimulus for development and an attraction for international investment would be favorable business regulations in the specialized zone free of most governmental frameworks such as taxation, customs, labor laws and other regulation parameters on business. The Kingdom of Saudi Arabia and the Saudi Arabian Public Investment Fund is backing the project with US$500 billion [6, 7]. One question I would raise would be availability of renewable water. This area of Saudi Arabia has 2 km^3 or 2 billion m^3 of renewable water in rivers that also feed aquifers in alluvial sediments that supply 10% of the country's drinking water. If a per capita allowance of 500 m^3 (a stress value) is planned, this would sustain 4 million inhabitants. The Saudi Arabian per capita renewable water was between 65 and 75 m^3 in 2014 and will be less in 2035 (Table 2.5). To make up the great water deficit, the Kingdom generates its drinking water needs from more than a score of desalination plants (50%) and confined (not renewable) aquifers (40%) as well as the 10% from renewable sources in the planned mega-city region. As has been stressed in previous chapters, the extraction of groundwater must be balanced by groundwater recharge lest the source of water be lost and/or detrimental subsidence affects the area. Finally, the effect of global warming on the water budget for the mega-city in future years, (e.g., 2035, 2050), based on updated Intergovernmental Panel on Climate Change reports, has to be factored in to the master plans for NEOM, especially the response of river sources from snow melt in the surrounding mountains.

8.3 Outside Funding for Mega-Projects to Rehabilitate Cities with Large Under-Serviced Populations

There are fews nations that can self finance such large scale comprehensive 'fix it' projects. Outside funding to complement national funding is the only way most nations can begin the rehabilitation/redevelopment of many major cities in order to serve their populations, thus enhancing their quality of life and hope for the future of their children.

As described in the previous section, India is financing the rehabilitation/redevelopment of important needs in existing cities with a 20 city pilot project and a budget of US$79 million each over 4 years (US$31, +US$16, +US$16, +US$16 million) from taxation mainly from the manufacturing, industrial and service sectors and input from the states and cities, and with contributions from the United States, Singapore, Spain, France, Germany, and Japan. Rehabilitation/redevelopment of existing major cities with grave social, economic, and infrastructure problems can be, in some cases, as costly as building a new city, a cost that can not be borne by a nation itself during the time rehabilitation/redevelopment takes place. Thus, less developed and developing nations will require much external financing to complement available self-funding via low cost, long term loans, and grants from international banking institutions and from economically strong nations wanting to improve political relations, and/or cultivate access to a receiver nation's natural resources. This is the situation for many nations with highly populated cities and mega-cities in Asia, Africa, and for some in South America.

8.3.1 Sources of Funding for Development/Redevelopment to Reduce Poverty and Improve Living Standards for Underserved People

Nations with the task of finding funding to redo their highly populated cities so that they function more efficiency and bring services to underserved neighborhoods have to invest in experienced planning teams to prepare detailed plans of what can and will be done to improve living conditions for all citizens. The plan must include controls on finances to assure transparency that stunts corruption that in some countries is 'a way of life' for the corrupters and those who support them directly or by turning a blind eye to their activities for a price. The plan would be to assure the public of the many essentials to sustain a stable society including the already repeated safe water, access to adequate sanitation, health care, readily accessible transport systems, and education. As a result, this will better living and working conditions for an entire urban population. Finally, a proposal to lenders should have a schedule for repayment of loans that they take out. International and national institutional lenders that may support such proposals include the International Bank for Reconstruction and Development (IRBD, a division of the World Bank Group), and the following regional development institutions: European Bank For Reconstruction And Development, the recently established Asian Infrastructure Investment Bank, the African Development Bank, and the Inter-American Development Bank. National agencies such as the United States Agency for International Development created by President John F. Kennedy awards grants to support necessary and well planned development projects as do the EU Europe Aid Development and Cooperation Unit, the French Development Agency, Germany's Economic Cooperation and Development Unit, The Japan International Cooperation Agency, the Norwegian Agency International Development Program, the UK Department for International Development, and others. Development/cooperation agencies such as the United Nations Development Program including its Food and Agriculture Organization component may provide technical support in lieu of funding.

8.3.1.1 Alternate Funding Source through the World Bank Group

The International Finance Corporation (IFC) of the World Bank Group encourages private ventures to assist developing countries to reach sustainable growth. The IFC does this by financing investment with capital it makes available from international sources. Another division of the World Bank Group, the Multilateral Investment Guarantee Agency (MIGA) complements the IFA mission by encouraging direct foreign investment in developing countries with access to guarantees (risk insurance) to investors and lenders.

8.3.2 Qualifications for Loans/Grants

The eligibility for development loans and grants that are made and the types of loans or grants available by the IBRD and the International Development Association (IDA) of the World Bank Group is membership in the World Bank. The Bank has 189 member countries. Whether a development loan/grant originates from the IBRD or the IDA is determined by a country's per capita gross national income and falls into two categories. Lower Income Countries (LIC), the poorer ones, have per capita annual incomes of US$1005 or less. These can receive IDA loans/grants. Middle Income Countries (MIC), those in a better development stage, can apply for IBRD loans/grants. These are divided into Lower MICs with per capita annual incomes of US$1006 to US$3955 and Upper MICs with per capita annual incomes of US$3956 to US$12,235. These figures change as the global economic status changes. Those given above are for 2016. Other World Bank members are not eligible for loans/grants. Seventy-nine countries are eligible for IBRD lending whereas 64 countries are eligible for IDA lending. There are 17 countries that can receive loans that are a blend of the two granting divisions. Conditions for a loan vary. The IDA loans to the LIC countries are given at zero interest for 40 years with a 10 year grace period. In general, the IBRD loans have an interest rate slightly above the market rate and are to be repaid in 12–15 years. It is likely that the development institutions listed in the previous paragraphs and others follow similar norms adjusted to their bylaws.

8.3.3 Grounds for Requesting Development Loans/Grants

The World Bank supports development projects that are sustainable and help to reduce poverty and improve the quality of life for impoverished citizens. In general, the lending institutions will provide technical assistance to government experts in the design and implementation of development projects. This implies negotiations that bring agreement on project primary and secondary objectives, agreement on productivity output as evaluated by performance measures, and finally agreement on stage implementation of a supported project. Table 8.1, modified from the World

Table 8.1 Stages in lending cycle from conception to implementation, completion, and post-completion evaluation

A country's proposal for obtaining a loan or grant for a development project that will be economically beneficial, reduce poverty, and improve the quality of life for its less advantaged citizens (e.g., extend clean water and sanitation networks)
Discussions with the lender on strategies to carry out the project including changes that improve the efficiency, consider consultant suggestions and technical assistance as needed, and put in place financial safeguards against corruption with the possibility that the project may be dropped at any time
Make proposal available to public for input from affected parties and experts not associated with the proposal
Preparation of the reworked proposal that includes monthly reports and reports as stages are completed
Negotiations with the lender on terms of a loan and of disbursement of funding as project stages are completed and new stages are ready to start as well as review of the need for additional technical assistance
Loan is approved so that project can be implemented under strict supervision and presentation of stage reports as agreed upon
Implementation to completion and a final status report
Post project completion evaluation to bring to light how efficiency can be improved for like future projects

Modified from World Bank lending project cycle [8, 9]

Bank, presents the sequence that could be followed in the preparation and submission of a development proposal that should also reduce poverty [8, 9]. As such, investments are in infrastructure for improving a clean water supply, access to adequate sanitation and waste disposal, electricity, roads, and urban regeneration/redevelopment.

Financial institutions are supportive of social development that can improve chances for long term increased productivity such as education and transfer of knowledge, healthcare services including nutrition advice, and agriculture via improved rural service advisors. Investments may support industrial projects that offer employment not only during construction but also after it and that result in additional employment in small businesses that supply goods and services to large projects. Investment in institutional development should bring about environmental protection regulation that prevents pollution that affects people and ecosystems, as well as stopping corruption, and insuring personal security through the law and an independent judicial system. In recent years attention is focusing on projects that show that they can readily adapt to global warming/climate change. All these aspects have to be considered in tailoring proposals for development submitted to specific lending institutions.

Proposals for financial assistance for rehabilitation/redevelopment of cities with high populations where many citizens live in poverty in shantytowns or slums underserved or not serviced by basics for a sustainable good quality of life will be judged in part by costs benefits analysis. This evaluation is based on assessments of short and long term economic data. In addition, the analysis should consider the long term social and environmental benefits, and capacity to adapt to new realities such as the effects of changing climate. A proposal must include as well a government's commitment to be responsible for the maintenance and sustainability of the rehabilitated/redeveloped urban systems that exist and those will be created with the requested financing.

8.4 Human and Other Natural Resources

Important factors to include in proposals for long term large scale projects include human resources, technical resources, and natural resources available to support the planned rehabilitation/redevelopment. The human resources without which planned projects could not be accomplished include manual laborers, heavy machinery operators, and experienced construction/reconstruction personnel (e.g., carpenters, roofers, electricians, plumbers, mechanics/repairmen, heating/air conditioning installers). An able workforce as given above and given in the following sentences should receive acceptable wages and benefits as befit their jobs. Basic to the reworking of an urban center are police to keep a city safe and moving during reconstruction and firefighters and medical personnel to respond to emergencies that may arise. Technical personnel from national colleges/universities that can contribute to a redevelopment proposal are scientists (geologists, biologists, chemists, environmentalists, and engineers [civil, code savvy, sanitary, mechanical, electrical, industrial, quality control, transportation]) and hazard risk assessors. The risk assessors should include adaptation protocols to mitigate existing and projected threats from global warming/climate change data as they may evolve during the century. The greater input there is from qualified, experienced, national and international experts will be positive for lender/grantor evaluation of a request for funding. This emphasizes the need to support educational institutions to be certain that people will be trained and educated to keep rehabilitated/redeveloped projects sustainable for a nation's future.

A nation's ability to provide natural resources for redevelopment construction and to sustain a good standard of living for all citizens when the job is done and for future populations that will settle in a city is important to proposal reviewers in their decision making. This would include wood, metal ores and industrial rocks and minerals and capacity to refine them into building materials (e.g., steel, cement, concrete, copper). Proposal reviewers will want to know the energy resources planned to service all users of a redone city (e.g., coal, oil, natural gas, nuclear, hydroelectric, solar, wind) and the ability to deliver energy to all as a city population increases. The goal of urban rehabilitation/redevelopment is to achieve full livability. As emphasized in past chapters, basic to sustainable livability are water security and food security, topics that have to be addressed in a proposal for financing a city rehabilitation project to demonstrate to evaluators that a growing city population can be sustained health wise and nutritionally.

References

1. Business Standard. (2015, July 9). Cabinet nod to Rs 1 lakh cr for urban renewal, 100 smart cities to take off.
2. Government of India. What is a smart city? Online. smartcities.gov.in/content/innerpage/what-is-smart-city.php click on 'read more'.
3. Moore, M., & Foster, P. (2011). China to create largest mega city in the world with 42 million people. The Telegraph (January 24). Online. www.telegraph.co.uk/news/worldnews/asia/china/8278315/China-to-create-largest-mega-city-in-the-world-with-42-million-people.html.
4. Wikipedia. (2015). Proposed new capital of Egypt. Online. www.en.wikipedia.org/wiki/Proposed_new_capital_of_Egypt
5. COHRE.(2008).Humanrightsorganizationscallforanendtohumanrights:Morethan800,000residentsevictedfromAbujafrom2003–2007violationsinNigeria.Online.www.reliefweb.int/reprint/nigeria/human-rights-organizations-call-end-housing-rights-violations-nigeria-more-800,000
6. NEOM. (2018). Frequently asked questions. Online. discoverneom.com/content/pdfs/NEOM_FAQS_ENGLISH.pdf
7. Al-Saud, M. (2010). Water sector of Saudi Arabia. Presented at the Japan Arabian economic forum, Tunis. Online. www.jccme.or.jp/english/jaef2_overview/meeting/session3/workshop2/18_w2.pdf
8. World Bank. (2017, July 18). World Bank project cycle. Online. www.worldbank.org/en/projects-operations/products_and_services/brief/projectcycle
9. Bretton Woods. (2004, July 26). The World Bank project cycle explained. Online. www.brettonwoodsproject.org/2004/07/art-62712/

Chapter 9
Planning for Contracting/Aging Populations: Societal and Economic Ramifications

9.1 Population Contraction

Many countries worldwide have below a population replacement birth rate (2.01) and/or an aging population. This has resulted in contracting populations with the degree of contraction varying between countries but resulting in similar predictable societal, economic, and to some degree political problems. Examples of countries with contracting populations are shown in Table 9.1.

9.2 Foreign/Guest Workers: A Solution?

With less young people available to enter a workforce and jobs to fill, there is the possibility of bringing in guest workers (foreign nationals). Those admitted preferentially fill skilled worker positions in manufacturing and industry while others are allowed in, ofttimes seasonally, to fill agricultural needs. Others serve as caregivers for the aged and as cleaning personnel, nannies and housekeepers to free up qualified women so that they can enter a country's workforce. Additional guest workers fill jobs that keep cities functioning such as store clerking, garbage collecting, and many other municipal jobs such as maintaining the infrastructure that is in place in most countries with contracting populations. As described below, the admission of guest workers to fill jobs such as those cited above is not new.

In 1942, the United States signed the Bracero Program with Mexico that allowed Mexican workers to enter the country to work in the agricultural sector for specific periods of time after which they returned to Mexico. This pact guaranteed the workers adequate living conditions such as sanitation, food, and shelter and a minimum wage (in 1942) of 30 cents an hour. This program was terminated in 1964. Subsequent legislation allowed employers to bring seasonal workers into the United States [2].

F. R. Siegel, *Cities and Mega-Cities*, SpringerBriefs in Geography,
https://doi.org/10.1007/978-3-319-93166-1_9

Table 9.1 Examples of countries with declining populations from 2017 projected to 2035 and 2050, in millions [1]

Country	Total fertility rate 2017	Population mid-2017	Population 2035	Population 2050	% Loss
Romania	1.2	19.6	17.4	13.9	29
Poland	1.4	38.4	36.9	32.6	15
Greece	1.3	10.7	10.4	9.1	15
Portugal	1.4	10.3	10.0	9.2	11
Spain	1.3	46.6	46.1	44.4	5
Japan	1.5	126.7	119.1	101.9	20
Ukraine	1.5	42.3	39.0	33.0	22
Bulgaria	1.5	7.1	6.4	5.3	18
Serbia	1.5	7.0	6.4	5.3	26
Cuba	1.7	11.3	10.8	9.8	13
Russia	1.7	146.8	147.9	144.8	0.1

During the 1960s, The Federal Republic of Germany had an exploding economy and many more jobs open than could be filled by German nationals in sectors such as construction, heavy industry, mining, and the automobile industry. The country allowed in workers from Italy, Greece and other Mediterranean nations to fill these jobs. In the 1980s, Turkish workers were brought in to fill municipal jobs. Today, 20% of Germany's population are of foreign origin and their German born off-spring. The conflicts in the Middle East during the second decade of the twenty-first century have brought many refugee asylum seekers to the European Union. Germany has received more than one million of the refugees and has been processing them in accordance with German immigration laws.

Japan is an industrialized country with a contracting population because of low birth rate and an aging society. By 2050, the Japanese population is projected to decrease by 20% (Table 9.1). Although Japan wished to maintain its ethnic character, it has had to import increasing numbers of guest workers. For example, in 2015 Japan had 29,000 foreign workers in construction. This number rose to 41,000 in 2016. In 2016, Japan had one million guest workers mainly from China (>30%), Vietnam (16%), and the Philippines (6%). These workers have a 3 year permit after which they must leave. Skilled workers have the possibility of residency after 5 or more years with Prime Minister Abe in 2017 promising to establish a fast track towards residency for them. Obviously, it is essential for guest workers to have at least a basic proficiency in Japanese to allow them to function without communication problems. Thus, guest worker applicants receive 400 hours of training in their home countries, 300 of which are in the Japanese language in which they must pass a test before receiving permission to move to Japan where they earn the yen equivalent of US$1500/month or more. It should be noted that guest workers contribute to a city/country tax base. To further acclimate into a society, it is important that guest workers in any country be well acquainted with the cultural norms of the country and its history in addition to their proficiency in the language. This knowledge can increase the possibility of permanent residency, permission to bring in immediate family, and the promise of future citizenship with all socio-economic benefits. In 2015, Japan published the 5th edition of its immigration control policy that recognizes the need to accept foreign nationals as guest workers given the country's contracting/aging population and the need to vitalize the economy and society and yet maintain the country's ethnic character [3]. It specifically defines what must be done by foreign nationals to achieve permanent resident status (such as taking on a Japanese name), and the conditions for acceptance of refugees. Naturalization is a lengthy process and no dual citizenship is permitted.

Resistance to the use of foreign nationals to fill critical and what some may think non-critical jobs would be self defeating. We can assess what could be the result of not having enough people, skilled/experienced and unskilled, in four sectors and their components: societal, economic, political, and environmental.

9.3 Meeting Societal Needs in an Aging/Contracting Population

We can first examine the implications of a contracting population on societal needs with fewer people to fill increasingly unfilled skilled and unskilled positions. Without enough medical doctors, nurses, and dentists, citizens are at risk when they get sick or are injured. This can happen in both urban and rural settings in countries with growing or contracting populations. For example, Brazil (with a growing population) imports medical doctors from Cuba on guest worker visas to provide healthcare services for underserved populations, especially in outlying towns and villages. A portion of their salaries goes to the Cuban government. The United States, also with an increasing population, issues residency permits to foreign medical doctors once they have completed hospital residency requirements in the United States. The country does the same for qualified nurses. These professionals practice both in urban centers and in poorly served rural populations. We have seen the public health problems that can develop if there is an extended strike of workers that collect garbage and trash and move it to secure disposal sites. If nationals are not filling these or other unskilled jobs because there are other employment opportunities in a contracting population, or because the jobs are "beneath them", a municipality can follow national immigration laws and import foreign workers as did Germany that brought in Turkish citizens in the 1980s. Other countries do the same in agriculture to tend livestock, work in slaughter houses, and work in the fields, whether in the United States or the European Union (including Great Britain until Brexit is finalized). Without sufficient people to fill agricultural jobs in a contracting population, food production may diminish and costs to citizens increase, a supply demand reality that contributes to inflation.

The same is the case for not having a full cadre of skilled personnel for manufacturing and industrial needs as well as for maintenance and repair of such facilities.

As noted previously, Japan with its strongly contracting population had one million foreign workers in 2016 (mainly Asians from China, The Philippines, and Vietnam) with a great number in the manufacturing sector (33%), 13% in the sales sector, and 12% in the hospitality sector.

The need for qualified teachers in grade school, middle school, and high school affects stable and growing societies as well as those losing population. Education whether at a high school level or at a university level among others (e.g., tradesman apprenticeships) is the key to job opportunities. For the most part teachers at the high school level are educated within a nation. Universities, however, bring in educators to give specialized courses and to do meaningful theoretical and applicable research nationally with the expectation that results will help the host country and be transferable to the international community.

Fundamental to societies in contracting, stable, or growing populations is personal security. Are there enough qualified persons, physically strong and mentally stable, to enter police and fire academies, graduate, and provide personal security? In Washington, D.C., for example, there have been hundreds of vacancies to be filled for the police to have a full complement of officers. In countries with contracting populations overall, but possibly with increasing urban populations, the lack of well trained police and firefighter personnel can put personal security at risk. The same can be said for having an educated, well-trained defense force that can protect national borders. Population size defines a country's defense posture, or if attacked, strike back. Drones, air power, an effective cyber force, and technical advances that are being made (e.g., long range laser weapons) that can drive back an enemy from terrain. However, as those who serve or have served in the military, and those who read the history of war or follow contemporary conflicts in the media know, it is the ground troops that are necessary to capture terrain, hold it and win a battle. Without enough eligible age citizens to draft or recruits to fulfill the requirements of a country's active armed forces, there can be a national problem. One solution is to induce foreign nationals that immigrate to join the military with the promise of fast track to citizenship in a country (e.g., the United States) with all the benefits it offers. It should be emphasized that with a full or nearly full complement of skilled and unskilled workers, there can be a rapid and effective response when a natural disaster impacts a population, whether in a growing or contracting population or in an urban or rural settings.

As previously stated, it is essential for guest workers to have more than a basic proficiency of a country's language so that they can both understand employers and express themselves well at work and in their everyday lives. A potential host country should provide language training in a potential guest worker's home country. During the language classes and subsequent immersion experiences, the trainers can assess worker language proficiency and acceptability for a move to the host country. Obviously, classes to make workers knowledgeable about the social and cultural norms of a host country as well as a basic knowledge of its history will be of great value to guest workers to more easily integrate into its population. It should be reemphasized that guest workers that enter a country legally pay taxes to city and national coffers. Those that enter a country illegally (e.g., the United States) and hold jobs pay taxes as well but can not draw on all the same benefits that citizens or legal guest workers do.

9.4 Economic Implications for Declining Populations

We have already touched upon the effects of national declining populations. If there is not enough population growth that will provide man/woman power to fill important, perhaps critical jobs, two results will be experienced. First, productivity will fall in the manufacturing and industrial sectors with a possible drop off in product quality. Profits will fall so that this and fewer employees will mean less domestic spending and less taxes being collected overall to support government services and programs. Second, with fewer agricultural workers and less crops being harvested, there is again a lesser tax base coming in from farm employees to help finance municipal, state, and government activities, and from taxes on the sale of farm products. Perhaps more important, there will be a rise in prices for the typical "food basket" that can fuel inflation to above "acceptable" levels and perhaps cause social unrest. Robotics, such as being continually developed in Japan to try to reduce the lack of human presence, are directed to reduce the need for skilled workers in industry (e.g., automobile production) and to present human like figures to do service work (e.g., receptionists, cleaners).

Fewer workers employed in commerce that suffers from less domestic spending will contribute to less taxes being collected. The communication, maintenance and repair, and transport sectors need experienced people to service the public. With fewer of them there will be a lesser tax income and unhappy populations if the services people need are delayed because of a lack of personnel (e.g., electrician, plumber, cell phone and computer service people, automobile mechanics, clerks, care givers for elders, etc.). These future prospects for countries with aging/contracting populations should stimulate government incentives such as those cited earlier so married couples have larger families now to counter the aging factor in the future as well as encouraging immigration of people prepared to fill such open positions as cited in previous paragraphs. The problems that affect a country's economic status with declining population will likely be emphasized in its urban centers that have increasing populations (~4 billion in 2018 with a projection to rise to 6.9 billion in 2050). It is clear that the unfilled jobs problems in countries with contracting populations can find solution by allowing guest workers, suitably prepared in the language, respectful of the culture, and knowledgeable about the history of a host nation.

The aging and shrinking of populations in many European countries puts their economies (GDPs) at risk and hence can diminish their joint international influence in solving socio-economic-political problems worldwide. Obviously, those countries with larger and younger stable or growing populations such as England (66.2

million in 2017 to 71.7 million in 2035 to 77.7 million in 2050) and France (65 million in 2016 to 68.5 million in 2035 to 72.3 million in 2050) [1] will challenge the German industrial sector (perhaps suffering from a shrinking workforce) and fare better economically than most countries on the continent. In general, most European countries have an established infrastructure that can readily absorb an inflow of guest workers that will sustain a good quality of life for them temporally and subsequently if they attain permanent residency and become naturalized citizens. As will be discussed in the following section, bringing in workers from abroad with time of stay limitations but with certain incentives may help reverse the effects of contracting populations and hence maintain a nation's productivity and ability to provide services to the public.

9.5 Political Options to Cope with Socio-Economic Changes that Result from Aging and Contracting Populations

An elected government, municipal or national, or an autocratic regime that heads a country that has an aging population and a declining population (fertility rate less than replacement rate) plus an emigration loss, often of educated and skilled segments of its citizenry, has to consider options to maintain or increase productivity of its economic sectors. This includes keeping inflation in check by making sure that supply meets demand especially for foodstuffs (including safe water), utilities, and transport. A government has the societal obligation to meet the essential needs and guarantee the right of the growing aged segments of its citizenry. This includes providing healthcare and care-giver assistance to the citizens who have spent their earlier lives contributing to society by doing the necessary jobs and paying into a tax base. It has to provide personal and national security so that police/fire/disaster response organizations and military units are kept up to specific man/woman power levels and with access to equipment and other means to do their jobs.

The obvious option to fill needs that exist is to attract guest workers for limited periods (e.g., at least 3 years) at an honest salary and reasonable benefits and perhaps with the possibility of permanent residency and citizenship with all the benefits it conveys. As noted previously, this is the approach Japan has embarked upon and serves as a model for South Korea, another Asian nation with a contracting population. This is not an option for an autocratic, repressive regime because guest workers will not opt to work there so that population stabilization and growth in such cases will have to be from within. Ignoring present and future problems caused by changes in a national population, aging or not, is not an option lest inaction stimulates unrest in its citizenry and governments ultimately be subject to change whether by free elections or by protest/revolutionary actions.

9.5.1 Population Increase by Refugee Political Asylum Seekers

There is a difference between invited guest workers and refugees that enter a country en masse seeking political asylum and without a thorough vetting before permitting entry. This is the case of refugees from the twenty-first century wars in Syria, Iraq (Middle East), plus others from Afghanistan (Asia), and from African countries (economic refugees) that are seeking to enter the so called "rich" countries for employment opportunities. Once entered, they can receive benefits that far out weigh help, if any, they can receive in their home countries. Any great influx of refugees presents a dilemma that today is causing socio-political problems for the European Union. The European Union assigned quotas for their members of the number of refugees they should take in but some European Union countries such as Poland, Hungary, the Czech Republic, Romania, and Slovakia are refusing to accept the assigned quota. Germany, on the other hand, allowed in 890,000 refugees in 2015, 250,000 more in 2016, and 90,000 during the first half of 2017. A vetting of these immigrants was not thorough initially but subsequently was tightened as a result of some criminal and terrorists acts. Persons with histories that could not be verified or with questionable backgrounds and many classified as economic refugees are being rejected and deported.

An added problem to the influx of refugees is the need to resettle families and individuals, provide them with homes, funds to live on, medical and dental insurance, and psychiatric/psychologic care. It is necessary as well to provide language classes until the refugees reached a level that would allow them to function in society both daily and when entering the workforce or higher education institutions. Economically, Germany can accept the properly vetted political asylum people. Other countries such as Italy and Greece that have influxes of asylum seekers trying to travel through them to reach the "rich" European countries (e.g., Germany, France, the Netherlands, Belgium, England) but do not have the economic resources to fully care for them in transit. Thus, they have to solicit funds from the EU and the UN for that purpose. The problem of how to deal with the refugees seeking political asylum has led to the rise of nationalistic conservative political parties in some EU countries (e.g., Germany, Hungary, Poland, the Czech Republic, and the UK). This is the result of feared social changes and economic responsibilities that admitting political refugees could entail.

It is interesting to note that by accepting more than 1.25 million refugees, Germany has put off its population decline. This is in addition to an annual vetted quota of 200,000 immigrants admitted. By mid-2014, before the great war refugee migration, Germany had a population of 80.9 million people with projections for mid-2035 of 80 million and for mid-2050 of 76.2 million with a fertility rate of 1.4, well below the 2.01 replacement rate. By mid-2015 and with the immigration influx, the country's population, still with a fertility rate of 1.5, rose to 81.1 million people with projected populations of 81.1 million by mid-2035, and 76.4 million for mid-2050. By mid-2016, Germany's population, with a fertility rate still at 1.5, rose to

82.6 million people with projections of 83.3 million by mid-2035 and 81 million by mid-2050. This latter figure (81 million for mid-2050) exceeds by 9.5 million the projected mid-2050 figure reported in 2012 making the 9.5 million person gain in 7 years questionable for this author. The projected population increase for 2050 published in 2016, if redacted correctly as reported, is likely related to the immigration and to government incentives to child-bearing age German couples to have more than one child. These incentives include tax benefits, child care facilities, and extended maternity leave from employment, but with pay (or a high percentage of the pay), for the mother and for the father. An essential addition to these benefits (especially for Japan) is better affordable housing that is comfortable for a family with two or three children. This may stem population contraction and result in population stability or growth. This incentives tact, wholly or in part, is being adopted by many nations with declining populations in their efforts to stimulate growth. The mid-2017 German population was 83.1 million with projections of 84.3 million for 2035 and 83.2 million for 2050, again with a 1.5 fertility rate that may not account for refugee child births, suggests that refugee inflow and government incentives to have larger families is contributing to the country's projected stable population through 2050 [4]. The existing and future economic benefits of the increased population for Germany and other countries with contracting populations are more workers to fill available jobs, more domestic spending, and an improved tax base.

9.6 Preservation of Environmental/Ecosystem Health with Some National Populations in Decline But with Global and Urban Populations Increasing

Populations are declining in most European nations and in Asia (Japan, South Korea, Taiwan, and Singapore). However, this represents about 1.3 billion people (~18% of the global population) versus 6.2 billion people in nations with growing populations. These are mainly in Asia and Africa but also in Latin America and to some degree in the Middle East. Certainly, nations with markedly contracting populations that do not bring in guest workers to match the decline in numbers of workers will have less of a demand on natural resources (e.g., water and food, energy). They will generate less sewage, garbage and solid wastes, and regulations on manufacturing/industrial operations should minimize releases of air and effluent pollutants into the immediate and neighboring environments and the sensitive ecosystems that serve humans' natural resource needs. In addition, it may be necessary to downsize some of the minimally used existing infrastructure but it is essential to regularly and thoroughly maintain it in good condition (e.g., health facilities, sewage collection and treatment plants, waste pick up and disposal, roads and bridges) to put it in use again to accommodate possible future natural population growth or following a national decision to admit large numbers of guest workers. The relief of pressure on "local" environments and their ecosystems in nations with contracting populations drawing less on them allows for previous ecosystem disruptions to heal and their resources to be harvested in a sustainable manner.

9.7 Challenges to Confront: Reiteration Plus

However, the global picture is not "rosy" as I write in 2017/2018 or for the projected future. Yes, nations with populations in decline represent ~18% of the 2017 global population but overall the other 82% is fueling earth's population growth. As reiterated in previous sections of this book, the population pressures to draw on natural resources (e.g., water, food, energy sources, wood, metal ores) will intensify driven especially by aspirations of many nations to improve their economies through increased industrialization. Such is less the case in most nations with declining populations. This and other problems will be most severely felt in large population urban centers (e.g., cities of millions and mega-cities with >10 million inhabitants) both in countries' coastal regions and inland into mid-continent environments (e.g., from extreme weather systems and rising sea level). The effects of global warming and climate change will continue and likely intensify unless greenhouse gas emissions (especially CO_2) decline dramatically from an international commitment to put into practice or develop new technologies that will slow down and ultimately arrest the industrial sector gas (CO_2) emissions to the atmosphere and emissions from vehicular sources. The Paris Accord is a good start to limit CO_2 emissions. Unfortunately, the United States administration with decision makers who seems to be non-conversant with the negative impacts of global warming/climate change opted not to join with the other nations of the world in this venture. One would hope that upon reflection, national political forces will force a reversal of this decision. Fortunately, many of the states in the United States are following established norms to decrease their CO_2 inputs to the atmosphere that has been successful. From 2005 to 2015, the United States greenhouse gas emissions (dominated by CO_2, with methane, nitrous oxide, and florinated gases) have decreased by 10% [5]. Nations with contracting populations, stable populations, and growing populations often with industrial driven economies, are signees to the accord striving to reduce CO_2 emission to the atmosphere. In theory, CO_2 emissions reduction should be coupled with a 'rapid' stabilization of global population to a number that can be carried by the earth's life sustaining natural resources. In practice, this is not the case because births outnumber deaths annually and global population is projected to reach 11 million or more by 2100 unless there is a population collapse. This could be by a pandemic naturally, crop failures worldwide, extraordinary loss of lives by war and the unfettered use of weapons of mass destruction (e.g., nuclear, biologic), or by the asteroid Apophis with its diameter of 1100 ft. (333 m) and a thickness of 700 ft. (212 m) that has a 1 in 45,000 chance of impacting our planet when it nears the Earth in 2036.

As an important corollary to the global warming solution, it will be necessary for technology to deal with an increasing contribution of methane to the atmosphere. The warming is abetting the decay of peat bogs and tundra and the release of this potent greenhouse gas into the atmosphere. Another potentially damaging methane source that could arise in the future is the possible decomposition of methane clathrate, a chemical compound that underlies the oceans continental shelf (average depth of 200 m) in high latitude regions. The clathrate is present in a thick sequence of sedimentary rocks in zones that are presently stable. If the stability is compromised by ocean warming, the release of methane through the seafloor of the continental shelf would add fuel to the global warming process. Seawater, to a 75 m depth, has a temperature that has increased 0.44 °C from 1971 to 2010 [6]. Whether seawater warming in the future will reach the rocks underlying the continental shelf and provide enough heat energy to destabilize methane clathrate is a valid question. It should be noted that methane clathrate can be viewed as a huge source of natural gas that in the future could be tapped but an exploration/exploitation venture could result in a blowout that would contribute to the greenhouse gases atmospheric contents. Research into the prevention of methane releases or how to damp their effect is of great importance.

Whether in countries with contracting, stable, or expanding populations, the keys to a nation's survival from global warming/climate change effects are economic strength, an educated, trained cadre of citizens, and a well developed regulatory and physical infrastructure that can be adapted to damp the impacts of events that can kill and injure people and destroy or damage property. The events include rising sea level and extreme weather that may be more frequent, more intense, and of longer duration than before climate change affected earth processes. For example, hurricanes can drive storm surges inland causing flooding reinforced by torrential or long lasting rain. In addition, droughts cause diminished water supplies with consequences for food security. Food security is further affected by warming that shifts agricultural zones. The warming shift also extends the reaches of disease bearing vectors (e.g., mosquitos). Further, heat waves are dangerous to public health and has killed tens of thousands in Europe in recent years. An effort has been made to rank survivability by country presented by regional infographics (for Europe, North America, South America, Central America, Africa, Asia and the Middle East) [7]. The ten countries proposed as likely to suffer most from climate change have rapidly growing populations, are less developed, and lack the wealth, infrastructure, and sufficient knowledgeable cadre and are mainly in Africa (Chad, Eritrea, Burundi, Democratic Republic of the Congo, Central African Republic, Sudan, Niger, Guinea Bissau) plus Haiti and Afghanistan. Conversely, the ten countries likely to suffer least from climate change have slow growing or stable populations, are developed, and mostly in Europe (Norway, Sweden, Finland, Denmark, United Kingdom, Germany, Iceland) plus New Zealand, Australia, and the United States.

Another big environmental challenge brought on by human activities that cater to the needs of the general global population is to maintain a sustainable food supply. Globally, soil degradation is a major problem as has been encroachment by population expansion. The preservation of the soil ecosystem is a priority to assure food security so that the two factors just cited be arrested. As stressed previously in the text, this means regulations that minimize or prevent encroachment (e.g., by expanding populations and by exploiters of natural resources). It means limiting erosion (by poor farming practices, wind, flowing water), maintaining (replenishing) soil nutrients to support growth, preventing crop loss from weeds that rob nutrients from soil and pests that spoil crop quality and yield. Obviously, where pollution is a soil/crop degrading factor (by toxic emissions and effluents, human activities), it must be stopped or greatly reduced. The preservation of forests by regulation and enforcement that controls lumbering and prevents clear cutting for expanding agriculture and mining is essential to save this soil-biome ecosystem. It should be noted that 1/3 of the global food supply is wasted by being disposed of at dumps, mainly in industrial societies from homes, from markets, and from restaurants. To this end, France's President Macron has put into legislation in 2018 requirements for these sources to either use waste of foods by sending them for compost or to donate unused food to food banks. The latter is already done globally by food establishments in many countries.

Overfishing the ocean ecosystems (e.g., near surface and deeper water fisheries) diminishes global food security. This affects contracting societies (e.g., Japan) and those with rapidly growing populations (e.g., Bangladesh, Indonesia, Ghana). Fisheries need time to reproduce between harvesting [8–11]. This means holding fishing boats (fleets) to international assigned tonnage harvesting allotments of specific species. It means apprehending and severely punishing rogue fishing boat owners and crew members with the possibility of losing boats and being fined and perhaps incarcerated for breaking national/international laws. The fisheries problems are further exacerbated by the fact that food fish species in a specific temperature range habitat migrate to cooler water when higher temperature of their natural habitat from global warming exceeds their normal comfort range (e.g., in the northern hemisphere fish migrate to the north or to deeper water). This means that fishing boats have to follow the migration expending more time at sea and driving up the costs for fuel and crew if on a daily wage basis. It also means importantly that coastal villages dependent on food fish for their protein intake can see this food source ebb due to fish migration. Ocean ecosystems are suffering as well from pollution by waterway pollutants that discharge into them and by the increasing mass of wastes that are disposed of in the oceans. Since 2011, radioactive waste waters from the earthquake/tsunami Fukushima nuclear power plant disaster are polluting Pacific Ocean environments, a process that continues in 2018. Fish from waters with radioactivity signals are not sellable domestically or in export markets, hence hurting the Japanese fishing sector.

We discussed air pollution in a past chapter (See Chap. 3, Sect. 3.2.1) as to sources, the fact that it impacts nations with populations in decline as well as those in growth, and what is being done, and what further can be done to limit air pollution. This atmospheric problem is another global ecosystem that suffers from an increasing human population on Earth and the demands of more people for goods and services from manufacturing, industrial, and other sectors. In some countries with contracting, stable, or growing populations, this makes the atmosphere a source of pollution that has caused sickness, and in some instances death. Waterway pollution suffers from the same sources and poisons their ecosystems and the life inhabiting them (see Sect. 3.22).

Aquifers and waterways (lakes, rivers, estuaries, seas and oceans) are sensitive, fragile ecosystems fighting to maintain ecological equilibrium worldwide. Their equilibrium and their very existence in many regions is threatened predominantly by human activities. Healthy aqueous ecosystems have to be preserved by all nations, growing or declining in population. Disrupted, damaged aqueous ecosystems have to be given chances to heal. Nearly destroyed or seemingly dead water bodies have to be literally brought back to life. This is being done for the Aral Sea by Uzbekistan and Kazakhstan by reestablishing the river water discharge into the Aral Sea that was largely cut off by the then Soviet Union. This rejuvenation project is supported by funding from the World Bank. All water systems have to be used in metered volumes so as to remain sustainable by careful planning and management that will prevent or significantly minimize overuse (discharge > recharge) and its effects. Water use must be controlled so as to stem pollution from natural and industrial driven human activities. Likewise there must be stoppage of misuse (by overfishing, waste disposal), and by rerouting (of rivers) to sustain industries at the expense of crop productive ecosystems. Water is a natural resource that is basic to sustain life on earth that has been discussed one way or another throughout the text (e.g., Sects. 2.5, 3.2.2, 3.2.3, 3.3, 4.4.1, 4.6.5, 5.3, 6.3, 7.3.3, 7.4, 7.5 and Table 5.2.).

References

1. Population Reference Bureau. (2017). *World population data sheet*. Washington, DC: Population Reference Bureau.
2. Wikipedia. (2017). Bracero Program.
3. Ministry of Justice, Japan. (2015). Basic plan for immigration control (5th edition), 45 p. Online. www.immi.-moj.go.jp/seisaku/2015_kihonkeikaku_honbun_pamphlet_english.pdf
4. Population Reference Bureau. (2014, 2015, 2016, 2017). World population data sheets, Washington, DC.
5. U. S. Environmental Protection Agency. (2017). Inventory of U. S. greenhouse gas emissions and sinks: 1990–2015. Washington, DC, 29 p. Online. www.epa.gov/ghgemissions/inventory-us-greenhouse-gas-emissions-and-sinks-1990-2015
6. Intergovernmental Panel on Climate Change. (2013). Fifth assessment report. In *Climate change 2013. The physical science basis. Summary for policy makers*. Cambridge, NY: Cambridge University Press 29 p. Online. http://www.ipcc.ch/report/ar5/wgl/.
7. Brodwin, E, & Johnston, M. (2015). The countries most likely to survive climate change in one info graphic. *Business Insider*, Nov. 30. Online. www.businessinsider.com/countries-that-will-survive-climate-change-infographic
8. Worm, B., Hilborn, R., Baum, J. K., Branch, T. A., Collie, J. S., Costello, C., et al. (2009). Rebuilding global fisheries. *Science, 325*, 578–585.

9. Costello, C., Ovando, D., Hilborn, R., Gaines, S., Deschesnes, O., & Lester, S. (2012). Status and solutions for the world's unassessed fisheries. *Science, 338*, 517–520. https://doi.org/10.1120/science.1223389.
10. World Ocean Review. (2013). Living with the ocean. WOR2 – Overview. The future of fish – The fisheries of the future. In N. Gelpke, N. Behnam, & M. Visbeck (Eds.), *marabous gGmbH*. Hamburg, Germany.: 143 p. Online. oceanrep.geomar.de/22085/1/WOR2_english.pdf.
11. Costello, C., Ovando, D., et al. (2016). Global fishery prospects under contrasting management regimes. *Proceedings National Academy of Sciences, 113*, 5125–5129. https://doi.org/10.1073/pnas.1520420113.

Epilogue

What should be done given an assessment of ills in a city is the question a planning team starts with. What could be done is determined in great part by economic resources but with influence of human and natural resources available to carry out proposed changes. What would be done initially and in prioritized stages with phased in disbursement of funds requires a realistic appraisal of problems in existing cities of what most needs to be done to improve and sustain the quality of life for all citizens. The prioritization is especially important in cities/mega-cities where shantytowns or slums that can have hundreds of thousands to millions of inhabitants have taken root without access to the necessities that allow people to have good health (e.g., clean water, ample nutrition, sanitation/waste collection, shelter), good services (e.g., electricity, gas, telecommunications, Wi-Fi), an educational lifeline for children (e.g., free basic and advanced schooling including apprenticeship opportunities), and a strong sense of self esteem (e.g., employment and its rewards).

The growing global population is projected to concentrate in urban centers (human ecosystems) so that by 2050, 6.9 billion people will inhabit cities with 2.9 billion in rural centers, versus the 2018 figures that have ~4 billion in cities, a 70% increase in 33 years. The loss in the rural population from 3.5 billion in 2018 to 2.9 billion in 2050 in rural areas represents a 14% change [1]. As the urbanization process progresses, environments and the ecosystems that comprise them will be disrupted by encroachment onto land by squatters creating densely populated shantytowns. Farmlands and other productive areas including waterways proximate to city boundaries will be further hurt by the wastes squatters create thereby diminishing the natural resource productivity of the invaded lands. As noted above, many municipal governments in developing and less developed countries (e.g., in Africa and Asia) have major cities with a high density of population including millions that do not have access to safe water, to adequate sanitation, and health clinics to serve the needs of scores of millions of inhabitants. We know that in 2017-2018, we are not sustaining large segments of the world's population, perhaps one billion souls, especially in urban settings, with an acceptable standard of living. If this situation is

F. R. Siegel, *Cities and Mega-Cities*, SpringerBriefs in Geography,
https://doi.org/10.1007/978-3-319-93166-1

not dealt with starting "now" to show marked improvements in the near future, the urbanization problems and refugees seeking asylum faced by many cities and nations today will intensify. The calls for changes in government politics/policies to address urban and country population ills will be greater in the near and foreseeable future. The fact is that we are not sustaining the global population as of 2018. The question is then can we sustain a projected 2050 global population with 2.3 billion more people or an additional 1.2+ billion in 2100 with ~70% inhabiting cities, many densely populated. Can we nourish them with clean water and food, with permanent shelter, with adequate sanitation, with health clinics, education, with employment opportunities, and with other services that well governed, well financed cities provide for their citizens? What are your thoughts? What world might your offspring inherit?

Reference

1. Population Reference Bureau. (2017). World population data sheet 2017. Washington, DC.

Index

F. R. Siegel, *Cities and Mega-Cities*, SpringerBriefs in Geography,
https://doi.org/10.1007/978-3-319-93166-1